PHYSICS AND CHEMISTRY OF CLASSICAL MATERIALS

Applied Research and Concepts

PHYSICS AND CHEMISTRY OF CLASSICAL MATERIALS

Applied Research and Concepts

Edited by
Ewa Kłodzińska, PhD

A. K. Haghi, PhD, and Gennady E. Zaikov, DSc
Reviewers and Advisory Board Members

Apple Academic Press

TORONTO   NEW JERSEY

Apple Academic Press Inc. | Apple Academic Press Inc.
3333 Mistwell Crescent | 9 Spinnaker Way
Oakville, ON L6L 0A2 | Waretown, NJ 08758
Canada | USA

©2015 by Apple Academic Press, Inc.

First issued in paperback 2021

Exclusive worldwide distribution by CRC Press, a member of Taylor & Francis Group
No claim to original U.S. Government works

ISBN 13: 978-1-77463-361-8 (pbk)
ISBN 13: 978-1-77188-045-9 (hbk)

Library of Congress Control Number: 2014953149

Library and Archives Canada Cataloguing in Publication

Physics and chemistry of classical materials: applied research and concepts / edited by Ewa Kłodzińska, PhD; A.K. Haghi, PhD, and G.E. Zaikov, DSc, Reviewers and Advisory Board Members.

Includes bibliographical references and index.
ISBN 978-1-77188-045-9 (bound)
1. Materials. 2. Materials--Analysis. 3. Materials science. 4. Chemistry, Technical.
I. Kłodzińska, Ewa, editor

TA404.2.P59 2014   620.1'1   C2014-906937-5

Apple Academic Press also publishes its books in a variety of electronic formats. Some content that appears in print may not be available in electronic format. For information about Apple Academic Press products, visit our website at **www.appleacademicpress.com** and the CRC Press website at **www.crcpress.com**

ABOUT THE EDITOR

Ewa Kłodzińska, PhD

Ewa Kłodzińska, PhD, is working at the Institute for Engineering of Polymer Materials and Dyes in Torun, Poland, and investigates surface characteristics of biodegradable polymer material on the basis of zeta potential measurements. She performed research on determination and identification of microorganisms using the electromigration techniques for the purposes of medical diagnosis for 10 years. She he has written several original articles, monographs, and chapters in books for graduate students and scientists and has made valuable contributions to the theory and practice of electromigration techniques, chromatography, sample preparation, and application of separation science in pharmaceutical and medical analysis. Dr. Ewa Kłodzińska is a member of editorial boards of *ISRN Analytical Chemistry* and the *International Journal of Chemoinformatics and Chemical Engineering* (IJCCE). She earned her PhD degree from Nicolaus Copernicus University, Faculty of Chemistry in Torun, Poland.

REVIEWERS AND ADVISORY BOARD MEMBERS

A. K. Haghi, PhD

A. K. Haghi, PhD, holds a BSc in urban and environmental engineering from University of North Carolina (USA); a MSc in mechanical engineering from North Carolina A&T State University (USA); a DEA in applied mechanics, acoustics and materials from Université de Technologie de Compiègne (France); and a PhD in engineering sciences from Université de Franche-Comté (France). He is the author and editor of 65 books as well as 1000 published papers in various journals and conference proceedings. Dr. Haghi has received several grants, consulted for a number of major corporations, and is a frequent speaker to national and international audiences. Since 1983, he served as a professor at several universities. He is currently Editor-in-Chief of the *International Journal of Chemoinformatics and Chemical Engineering* and *Polymers Research Journal* and on the editorial boards of many international journals. He is a member of the Canadian Research and Development Center of Sciences and Cultures (CRDCSC), Montreal, Quebec, Canada.

Gennady E. Zaikov, DSc

Gennady E. Zaikov, DSc, is Head of the Polymer Division at the N. M. Emanuel Institute of Biochemical Physics, Russian Academy of Sciences, Moscow, Russia, and Professor at Moscow State Academy of Fine Chemical Technology, Russia, as well as Professor at Kazan National Research Technological University, Kazan, Russia. He is also a prolific author, researcher, and lecturer. He has received several awards for his work, including the Russian Federation Scholarship for Outstanding Scientists. He has been a member of many professional organizations and on the editorial boards of many international science journals.

CONTENTS

LIST OF CONTRIBUTORS

Arezoo Afzali
University of Guilan, Rasht, Iran

Korablev Grigory Andreevich
Doctor of Chemical Science, Professor, Head of the Laboratory of Basic Research and Educational Center, Udmurt Scientific Center, Ural Division, Russian Academy of Science; Head of the Department of Physics at Izhevsk State Agricultural Academy. 426052, Izhevsk, 30 let Pobedy St., 98-14, Tel.: +7 (3412) 591946; E-mail: biakaa@mail.ru, korablevga@mail.ru

V. A. Babkin
Volgograd State Architect-Build University Sebrykov Departament, 403300 Michurin Street 21, Michailovka, Volgograd region, Russia, E-mail: sfi@reg.avtlg.ru

Yu. V. Berestneva
Donetsk National University, 24 Universitetskaya Street, 83 055 Donetsk, Ukraine, E-mail: chembio@sky.chph.ras.ru

S. N. Bondarenko
Volzhsky Polytechnical Institute (branch) Volgograd State Technical University 42a Engelsa Street, Volzhsky, Volgograd Region, 404121, Russian Federation, E-mail: vtp@volpi.ru; www.volpi.ru

Rustam Ya Deberdeev
Kazan National Research Technological University, 68 Karl Marx Street, 420015 Kazan, Republic of Tatarstan, Russian Federation, Fax: +7 (843) 231-41-56; E-mail: n.v.ulitin@mail.ru

K. S. Dibirova
Dagestan State Pedagogical University, Makhachkala 367003, Yaragskii st., 57, Russian Federation

V. U. Dmitriev
Volgograd State Architect-Build University Sebrykov Departament, 403300 Michurin Street 21, Michailovka, Volgograd region, Russia, E-mail: sfi@reg.avtlg.ru

I. V. Dolbin
Kh.M. Berbekov Kabardino-Balkarian State University, KBR, Nal'chik 360004, Chernyshevskii st., 173, Russian Federation

Alexei V. Dubrovsky
Institute of Theoretical and Experimental Biophysics RAS, Institutskaya Street 3, 142290 Pushchino, Moscow region, Russia

M. A. Esteso
U.D. Química Física, Universidad de Alcalá. 28871 Alcalá de Henares (Madrid), Spain, E-mail: miguel.esteso@uah.es, carmen.teijeiro@uah.es.

A. N. Garashchenko
Volzhsky Polytechnical Institute (branch) Volgograd State Technical University 42a Engelsa Street, Volzhsky, Volgograd Region, 404121, Russian Federation, E-mail: vtp@volpi.ru; www.volpi.ru

K. Z. Gumargalieva
N. N. Semenov Institute of Chemical Physics, RAS, 4 Kosygin Street

H. Hlídková
Institute of Macromolecular Chemistry Academy of Sciences of the Czech Republic, v.v.i. Heyrovský Sq. 2, 162 06 Prague 6, Czech Republic

D. Horák
Institute of Macromolecular Chemistry Academy of Sciences of the Czech Republic, v.v.i. Heyrovský Sq. 2, 162 06 Prague 6, Czech Republic, E-mail: horak@imc.cas.cz

M. N. Ignatov
Mechanical Engineering Faculty, National Research Perm Technical University, Perm, Russia, E-mail: ignatovaanna2007@rambler.ru

A. M. Ignatova
Mechanical Engineering Faculty, National Research Perm Technical University, Perm, Russia, E-mail: ignatovaanna2007@rambler.ru

A. N. Inozemtsev
M. V. Lomonosov MSU, Biological Faculty, Leninskie Gory, 119991 Moscow, Russia, E-mail: olga-karp@newmail.ru.

V. F. Kablov
Volzhsky Polytechnical Institute (branch) Volgograd State Technical University 42a Engelsa Street, Volzhsky, Volgograd Region, 404121, Russian Federation, E-mail: vtp@volpi.ru; www.volpi.ru

O. V. Karpukhina
N. N. Semenov Institute of Chemical Physics, RAS, 4 Kosygin Street

L. I. Kazakova
Institute of Theoretical and Experimental Biophysics Russian Academy of Science, 142290 Pushchino, Moscow region 142290, Russia

N. A. Keibal
Volzhsky Polytechnical Institute (branch) Volgograd State Technical University 42a Engelsa Street, Volzhsky, Volgograd Region, 404121, Russian Federation, E-mail: vtp@volpi.ru; www.volpi.ru

I. A. Kirsh
Moscow National University of Food Production, Moscow, Russia, E-mail: kaf.vms@rambler.ru

S. V. Kolesov
Bashkir State University, Russia, Republic of Bashkortostan Ufa-450074, ul Zaki Validi, 32, Tel:+7 (347) 229 96 14, E-mail: alenakulish@rambler.ru, The Institute of Organic Chemistry of the Ufa Scientific Centre the Russian Academy of Science, October Prospect 71, 450054 Ufa, Russia

G. V. Kozlov
N. M. Emanuel Institute of Biochemical Physics of Russian Academy of Sciences, Moscow 119334, Kosygin st., 4, Russian Federation / Dagestan State Pedagogical University, Makhachkala 367003, Yaragskii st., 57, Russian Federation, Kh.M. Berbekov Kabardino-Balkarian State University, KBR, Nal'chik 360004, Chernyshevskii st., 173, Russian Federation

T. V. Krekaleva
Volzhsky Polytechnical Institute (branch) Volgograd State Technical University 42a Engelsa Street, Volzhsky, Volgograd Region, 404121, Russian Federation, E-mail: vtp@volpi.ru; www.volpi.ru

E. I. Kulish
Bashkir State University, Russia, Republic of Bashkortostan Ufa 450074, ul Zaki Validi, 32, Tel.: +7 (347) 229 96 14; E-mail: alenakulish@rambler.ru, The Institute of Organic Chemistry of the Ufa Scientific Centre the Russian Academy of Science, October Prospect 71, 450054 Ufa, Russia

M. S. Lobanova
Volzhsky Polytechnical Institute (branch) Volgograd State Technical University 42a Engelsa Street, Volzhsky, Volgograd Region, 404121, Russian Federation, E-mail: vtp@volpi.ru; www.volpi.ru

Shima Maghsoodlou
University of Guilan, Rasht, Iran

G. M. Magomedov
Dagestan State Pedagogical University, Makhachkala 367003, Yaragskii st., 57, Russian Federation.

A. K. Mikitaev
Kh. M. Berbekov Kabardino-Balkarian State University, KBR, Nal'chik 360004, Chernyshevskii st., 173, Russian Federation

V. Ulitin Nikolai
Kazan National Research Technological University, 68 Karl Marx Street, 420015 Kazan, Republic of Tatarstan, Russian Federation, Fax: +7 (843) 231-41-56; E-mail: n.v.ulitin@mail.ru

Alexey V. Oparkin
Kazan National Research Technological University, 68 Karl Marx Street, 420015 Kazan, Republic of Tatarstan, Russian Federation, Fax: +7 (843) 231-41-56; E-mail: n.v.ulitin@mail.ru

I. A. Opeida
L. M. Litvinenko Institute of Physical Organic and Coal Chemistry National Academy of Sciences of Ukraine. 70 R. Luxemburg Street, 83 114, Donetsk, Ukraine

E. N. Pasternak
Donetsk National University, 24 Universitetskaya Street, 83 055 Donetsk, Ukraine

D. A. Provotorova
Volzhsky Polytechnical Institute, branch of Federal State Budgetary Educational Institution of Higher Professional Education "Volgograd State Technical University" 42a Engels Str., 404121, Volzhsky, Volgograd Region, Russia, E-mail: d.provotorova@gmail.com; www.volpi.ru

E. V. Raksha
Donetsk National University, 24 Universitetskaya Street, 83 055 Donetsk, Ukraine

Ana C. F. Ribeiro
Department of Chemistry, University of Coimbra, 3004-535 Coimbra, Portugal, Tel: +351-239-854460; Fax: +351-239-827703, E-mail: anacfrib@ci.uc.pt.

Ekaterina A. Saburova
Institute of Theoretical and Experimental Biophysics RAS, Institutskaya Street 3, 142290 Pushchino, Moscow region, Russia

G. A. Savin
Volgograd State Pedagogical University, 40013, Lenin Street, 27, Volgograd, Russia E-mail: gasavin@mail.ru

E. O. Sazonova
Moscow National University of Food Production, Moscow, Russia, post-graduate student E-mail: kaf.vms@rambler.ru

L. I. Shabarchina
Institute of Theoretical and Experimental Biophysics Russian Academy of Science, 142290 Pushchino, Moscow region 142290, Russia, E-mail: shabarchina@rambler.ru

Lyudmila I. Shabarchina
Institute of Theoretical and Experimental Biophysics RAS, Institutskaya Street 3, 142290 Pushchino, Moscow region, Russia, E-mail: tikhonenkosa@gmail.com

Evgenii B. Shirokih
Kazan National Research Technological University, 68 Karl Marx Street, 420015 Kazan, Republic of Tatarstan, Russian Federation, Fax: +7 (843) 231-41-56; E-mail: n.v.ulitin@mail.ru

A. S. Shurshina
Bashkir State University, Russia, Republic of Bashkortostan, Ufa, 450074, ul Zaki Validi, 32, Tel: +7 (347) 229 96 14, E-mail: alenakulish@rambler.ru, The Institute of Organic Chemistry of the Ufa Scientific Centre the Russian Academy of Science, October Prospect 71, 450054 Ufa, Russia

Abílio J. F. N. Sobral
Department of Chemistry, University of Coimbra, 3004-535 Coimbra, Portugal, Tel: +351-239-854460; Fax: +351-239-827703, E-mail: asobral@ci.uc.pt

A. G. Stepanova
Volzhsky Polytechnical Institute (branch) Volgograd State Technical University 42a Engelsa Street, Volzhsky, Volgograd Region, 404121, Russian Federation, E-mail: vtp@volpi.ru; www.volpi.ru

G. B. Sukhorukov
Institute of Theoretical and Experimental Biophysics Russian Academy of Science, 142290 Pushchino, Moscow region 142290, Russia, School of Engineering AND Materials Science, Queen Mary University of London, London, UK

Carmen Teijeiro
U.D. Química Física, Universidad de Alcalá. 28871 Alcalá de Henares (Madrid), Spain, E-mail: miguel.esteso@uah.es, carmen.teijeiro@uah.es

Sergey A. Tikhonenko
Institute of Theoretical and Experimental Biophysics RAS, Institutskaya Street 3, 142290 Pushchino, Moscow region, Russia, E-mail: tikhonenkosa@gmail.com

E.S. Titova
Volgograd State Technical University, 40013, Lenin Street, 28, Volgograd, Russia

Luís M. P. Veríssimo
Department of Chemistry, University of Coimbra, 3004-535 Coimbra, Portugal, Tel: +351-239-854460; Fax: +351-239-827703, E-mail: luisve@gmail.com

Gennady Efremovich Zaikov
Doctor of Chemical Science, Professor of N.M. Emanuel Institute of Biochemical Physics, RAS, Russia, Moscow, 119991, 4 Kasygin St., Tel.: +7 (495) 9397320, E-mail: chembio@sky.chph.ras.ru

A. A. Zhivaev
Volzhsky Polytechnical Institute (branch) Volgograd State Technical University 42a Engelsa Street, Volzhsky, Volgograd Region, 404121, Russian Federation, E-mail: vtp@volpi.ru; www.volpi.ru

N. A. Turovskij
Donetsk National University, 24 Universitetskaya Street, 83 055 Donetsk, Ukraine, E-mail: N.Turovskij@donnu.edu.ua

LIST OF ABBREVIATIONS

AIBN	Asobis (Isobutirnitrile)
AM	Antibiotic Amikacin
BPO	Benzoyl Peroxide
ChT	Chitosan
ChTA	Chitosan Acetate
CNR	Chlorinated Natural Rubber
CPM	Conductive Polymer Materials
CXR	Cyclohexane Regain
DBTC	Di Benzyl Tritio Carbonate
EDMA	Ethylene Dimethacrylate
EDTA	Ethylene Diamine Tetra Acetic Acid
FOM	Federation Object Model
HDPE	High Density Polyethylene
HEMA	Hydroxyethyl Methacrylate
HLA	High Level Architecture
HVSEM	High-Vacuum Scanning Electron Microscopy
LbL	Layer-by-Layer
LLDPE	Linear Low Density Polyethylene
LVSEM	Low-Vacuum Scanning Electron Microscopy
MC	Monte Carlo
MD	Molecular Dynamics
MMD	Molecular-Mass Distribution
MMT	Montmorillonite
MS	Medicinal Substance
OMT	Object Model Template
OOP	Object-Oriented Programming
PAH	Poly (Allylamine Hydrochloride)
PHEMA	Poly (2-hydroxyethyl methacrylate)
PMC	Polyelectrolyte Microcapsules
PPMI	Poly Pyro Meltitimide
PSS	Poly (Styrene Sulfonate)
PSt	Sulfonated polystyrene
PVC	Poly (Vinyl Chloride)
RSA	Rheometric Scientific Instrument
RTI	Runtime Infrastructure

RWFT	Random Walks in Fractal Time
SEM	Scanning Electron Microscopy
SOM	Simulation Object Model
TC	Technical Carbon

LIST OF SYMBOLS

C_M	conversion of monomer
d	dimension of Euclidean space
D	diffusion coefficient of C in austenite
df	fractal dimension
ds	spectral dimension
E	elasticity modulus
En	yield stress
E0	general energy
Eel	electronic energy
f	volume fraction of precipitates
G'	grow rate
I_s	nucleation rate at unit area
Ii	energy correspond for particular pixel
Imean	average image energy of ROI areas
l0	chain skeletal bond length
l0	length of the main chain skeletal bond
lcom	polymeric material burning depth
K	crystallinity degree
Me	molecular weight
m0	initial mass
M	mobility of interface
N	nucleation rate for unit volume
NA	Avogadro number
Npixel	total number of pixels
ncl	nano cluster
η_0	number of austenite grains at a unit volume (dissolved situation)
R'	outgoing group
r	tip-radius of growing phase

r	average size of precipitates
S	cross-sectional area of macromolecule
T	temperature
t	time
V_m	molar volume
wd	weight of dry sample
Wn	organoclay mass content
ww	weight of hydrated sample
X_e, Y_e	extended volume fraction
X_dyn	fraction dynamically recrystallized
x	fractions of elements
X, Y	actual volume fraction
Z	stabilizing group

Greek Symbols

ν	Poisson's ratio
χ	relative fraction
αam	amorphous phase
vas	antiphase
vF	Flory exponent
Δm	weight the absorbed film
ρn	density
ρp	polymer density
vs	in-phase
σY	yield stress
ΔG_m	transformation from austenite to ferrite
d_γ	austenite grain size after reheating of slab
ΔP	pressure change in the system
α	ferrite
γ	austenite
μ	chemical potential
τ	time when the new phase nucleates at plane B

PREFACE

This book provides a comprehensive presentation of the concepts, properties, and applications of classical materials. It also provides the first unified treatment for the broad subject of classical materials. Authors of each chapter use a fundamental approach to define the structure and properties of a wide range of solids on the basis of the local chemical bonding and atomic order present in the material. Emphasizing the physical and chemical origins of different material properties, this important volume focuses on the most technologically important materials being used and developed by scientists and engineers.

This new book is a collection of chapters that highlight some important areas of current interest in polymer products and chemical processes. This book

- focuses on topics with more advanced methods;
- emphasizes precise mathematical development and actual experimental details;
- analyzes theories to formulate and prove the physicochemical principles;
- provides an up-to-date and thorough exposition of the present state-of-the-art of complex materials;
- familiarizes the reader with new aspects of the techniques used in the examination of polymers, including chemical, physicochemical and purely physical methods of examination;
- describes the types of techniques now available to the chemist and technician and discusses their capabilities, limitations, and applications.

This book presents peer-reviewed chapters and survey articles on review, research, and development in the fields of classical materials and offers scope for academics, researchers, and engineering professionals to present their research and development works that have potential for applications in several disciplines of classical materials. The wide coverage makes this book an excellent reference book for researchers and graduate students on the subject. The new topics covered in this book will be an excellent resource for industries and academic researchers as well.

CHAPTER 1

A LECTURE NOTE ON QUANTUM-CHEMICAL CALCULATION OF MOLECULE OF 1-[2-(O-ACETYLMETHYL)-3-O-ACETYL-2-ETHYL]-METHYLDICHLORINEPHOSPHITE (I) AND COMPONENTS OF ITS SYNTHESIS

V. A. BABKIN, V. U. DMITRIEV, G. A. SAVIN, E. S. TITOVA, and G. E. ZAIKOV

CONTENTS

ABSTRACT

Quantum-chemical calculation of molecule of 1-[2-(o-acetylmethyl)-3-o-acetyl-2-ethyl]-methyldichlorinephosphite (I) and components of its synthesis [4] for the first time is executed by method AB INITIO in basis 6-311G**. The optimized geometrical and electronic structures of these compounds are received. Their acid force is theorized. All these compounds concern to a class of very weak C–H–acids (pKa > 14, where pKa is the universal parameter of acidity). Dependence between acid force of components of synthesis pKa (II) and pKa (III) and acid force pKa (I)

a required product is positioned: $pKa(I) = pKa\ (II) + \sqrt{\dfrac{pKa(III)}{4}}$

1.1 AIMS AND BACKGROUNDS

1-[2-(o-Acetylmethyl)-3-o-acetyl-2-ethyl]-methyldichlorinephosphite (I) is intermediate substance for reception of a medical product from hepatitis. It possesses interesting and probably unique properties.

This compound is received at interaction of acetyl chloride (III) and 5-acetyloxymethyl-2-chlorine-5-ethyl-1,2,3-dioxaphosphorynan (II) in a gas phase:

$$\text{I} \qquad\qquad\qquad \text{II} \qquad\qquad\qquad \text{III}$$

The Mechanism of reaction of acylation of bicyclophosphites by chlorine anhydride of carboxylic acids consists of three stages. The first stage of reaction is investigated in Ref. [4]. The mechanism of the second stage of reaction now is not studied. One of the first investigation phases of the mechanism of synthesis of studied compound (I) can consider an estimation of acid force of components of synthesis pKa (II) and pKa (III) and a required product pKa (I) and an establishment of dependence between them.

The purpose of the chapter was quantum-chemical calculation of components of synthesis of acetyl chloride (III) and 5-acetyloxymethyl-2-chlorine-5-ethyl-1,2,3-dioxaphosphorynan (II) and a product 1-[2-(o-acetylmethyl)-3-o-acetyl-2-ethyl]-methyldichlorinephosphite (I) by method AB INITIO in basis 6-311G**, a theoretical estimation of their acid force and an establishment of dependence between acid force of components of synthesis pKa (III) and pKa (II) and pKa (I).

1.2   EXPERIMENTAL PART

Method AB INITIO in basis 6-311G** with optimization of geometry on all parameters by the standard gradient method which has been built in PC GAMESS, used for quantum-chemical calculation of components of synthesis of 1-[2-(o-acetylmethyl)-3-o-acetyl-2-ethyl]-methyldichlorinephosphite (I) [1]. Calculation was carried out in approach of the isolated molecule a gas phase. The theoretical estimation of acid force of components of synthesis was carried out under the formula [2]:

$$pKa = 49.04 - 134.61 \times q_{MAX}^{H+},$$

where pKa is a universal parameter of acidity, and q_{MAX}^{H+} is the maximal charge on atom of hydrogen of a molecule.

For visual representation of models of components of synthesis program Mas-MolPlt [3] was used.

1.3   RESULTS OF CALCULATIONS

The Optimized geometrical and electronic structures, the general energy, electronic energy, lengths of bonds and valent corners of product 1-[2-(o-acetylmethyl)-3-o-acetyl-2-ethyl]-methyldichlorinephosphite (I) are received by method AB INITIO in basis 6-311G** and presented on Fig. 1.1 and in Table 1.1. Values pKa of all components of synthesis are certain by means of the formula, which used with success in [5–24]:

$$pKa = 49.04 - 134.61 \times q_{MAX}^{H+}$$

where q_{MAX}^{H+} (I) = +0.14, q_{MAX}^{H+}(II) = +0.16 and q_{MAX}^{H+}(III) = +0.14. Accordingly, pKa (I) = 30.2, pKa (II) = 27.5 and pKa (III) = 30.1. By us it has been positioned, that between acid force of components of synthesis pKa (II) and pKa (III) and acid force of a received product pKa (I) there is a following dependence:

$$pKa(I) = pKa \text{ (II)} + \sqrt{\frac{pKa(III)}{4}}$$

The General energy (E_0), the electronic energy (E_{el}), the maximal charge on atom of hydrogen -q_{MAX}^{H+} and values pKa of components of synthesis pKa (II) and pKa (III) and a product pKa (I) are presented in Table 1.2.

Quantum-chemical calculation of a molecule 1-[2-(o-acetylmethyl)-3-o-acetyl-2-ethyl]-methyldichlorinephosphite (I) for the first time is executed by method AB INITIO in basis 6-311G**. Quantum-chemical calculation of components of synthesis (II) and (III) has been executed by us earlier [4].

The optimized geometrical and electronic structures of compound (I) are received. Acid force is theorized. By us it is shown, that all of these compounds concern to a class weak C-H-acids (pKa > 14). Dependence between acid force of components of synthesis (II) and (III) and also acid force of a product (I) is positioned.

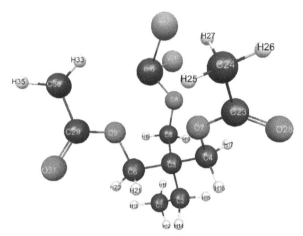

FIGURE 1.1   Geometric and electronic structure of a molecule of 1-[2-(o-acetylmethyl)-3-o-acetyl-2-ethyl]-methyldichlorinephosphite ($E_0 = -5,306,055$ kDg/mol, $E_{el} = -10,395,257$ kDg/mol).

TABLE 1.1   Optimized Lengths of Bonds, Valency Corners and Charges of Atoms of a Molecule of 1-[2-(o-acetylmethyl)-3-o-acetyl-2-ethyl]-methyldichlorinephosphite

Bond lengths	R, Å	Valency corners	Grad	Atom	Charge (by Milliken)
C(1)-C(2)	1.53			C1	−0.23
C(2)-C(3)	1.55	C(3)C(2)C(1)	117	C2	−0.21
C(3)-C(4)	1.53	C(4)C(3)C(2)	105	C3	−0.37
C(3)-C(5)	1.53	C(5)C(3)C(2)	108	C4	0.19
C(3)-C(6)	1.53	C(6)C(3)C(2)	107	C5	0.17
C(4)-O(7)	1.41	O(7)C(4)C(3)	111	C6	0.19
C(5)-O(8)	1.42	O(8)C(5)C(3)	111	O7	−0.44
C(6)-O(9)	1.41	O(9)C(6)C(3)	110	O8	−0.66
O(8)-P(10)	1.58	P(10)O(8)C(5)	121	O9	−0.45
C(1)-H(11)	1.08	H(11)C(1)C(2)	112	P10	0.85
C(1)-H(12)	1.08	H(12)C(1)C(2)	107	H11	0.09
C(1)-H(13)	1.08	H(13)C(1)C(2)	107	H12	0.11

TABLE 1.1 *(Continued)*

Bond lengths	R, Å	Valency corners	Grad	Atom	Charge (by Milliken)
C(2)-H(14)	1.08	H(14)C(2)C(3)	108	H13	0.10
C(2)-H(15)	1.08	H(15)C(2)H(14)	105	H14	0.11
C(4)-H(16)	1.08	H(16)C(4)O(7)	108	H15	0.11
C(4)-H(17)	1.08	H(17)C(4)H(16)	107	H16	0.11
C(5)-H(18)	1.08	H(18)C(5)O(8)	106	H17	0.12
C(5)-H(19)	1.08	H(19)C(5)H(18)	108	H18	0.13
C(6)-H(20)	1.08	H(20)C(6)O(9)	107	H19	0.11
C(6)-H(21)	1.08	H(21)C(6)H(20)	107	H20	0.11
P(10)-Cl(22)	2.06	Cl(22)P(10)O(8)	98	H21	0.12
O(7)-C(23)	1.32	C(23)O(7)C(4)	116	Cl22	−0.26
C(23)-C(24)	1.50	C(24)C(23)O(7)	111	C23	0.47
C(24)-H(25)	1.08	H(25)C(24)C(23)	109	C24	−0.25
C(24)-H(26)	1.08	H(26)C(24)C(25)	110	H25	0.12
C(24)-H(27)	1.08	H(27)C(24)C(25)	108	H26	0.13
C(23)-O(28)	1.18	O(28)C(23)O(7)	123	H27	0.14
O(9)-C(29)	1.32	C(29)O(9)C(6)	116	O28	−0.47
C(29)-C(30)	1.50	C(30)C(29)O(9)	111	C29	0.47
C(29)-O(31)	1.18	O(31)C(29)O(9)	123	C30	−0.26
P(10)-Cl(32)	2.08	Cl(32)P(10)O(8)	100	O31	−0.46
C(30)-H(33)	1.08	H(33)C(30)C(29)	109	Cl32	−0.29
C(30)-H(34)	1.08	H(34)C(30)H(33)	107	H33	0.13
C(30)-H(35)	1.08	H(35)C(30)H(34)	110	H34	0.13
				H35	0.13

TABLE 1.2 The General Energy, Energy of Bonds, the Maximal Charge on Atom of Hydrogen, a Universal Parameter of Acidity of Components of Synthesis of 1-[2-(o-acetylmethyl)-3-o-acetyl-2-ethyl]-methyl Dichlorinephosphite

№ п/п	The Component of synthesis	E_0, kDg/mol	E_{el}, kDg/mol	q_{MAX}^{H+}	pKa
1.	5-Acetyloxymethyl-2-chlorine-5-ethyl-1,2,3-dioxaphosphorynan	−3,700,998	−7,080,777	+0.16	27.5
2.	1-[2-(o-Acetylmethyl)-3-o-acetyl-2-ethyl]-methyldichlorinephosphite	−5,306,055	−10,395,257	+0.14	30.2
3.	Acetyl chloride	−1,604,983	−1,998,646	+0.14	30.1

KEYWORDS

- 1-[2-(o-acetylmethyl)-3-o-acetyl-2-ethyl]-methyldichlorinephosphite
- 5-acetyloxymethyl-2-chlorine-5-ethyl-1,2,3-dioxaphosphorynan
- Acetyl chloride
- Acidic force
- Method AB initio
- Quantum-chemical calculation

REFERENCES

1. Schmidt, M. W., Baldrosge, K. K., Elbert, J. A., Gordon, M. S., Enseh, J. H., Koseki, S., Matsvnaga, N., Nguyen, K. A., Su, S. J. et al. (1993). *J. Computer Chem., 14,* 1347–1363.
2. Babkin, V. A., Fedunov, R. G., Minsker, K. S. et al. (2002). *Oxidation Communication, 1(25),* 21–47.
3. Bode, B. M., & Gordon, M. S. (1998). *J. Mol. Graphics Mod., 16,* 133–138.
4. Babkin, V. A., Dmitriev, V. U., Savin, G. A., & Zaikov, G. E. (2009). Estimation of Acid Force of Components of Synthesis of 5–acetyloxymethyl-2-Chlorine-5-ethyl-1,2,3-dioxaphosphorynan Moscow, *Encyclopedia of the Engineer Chemist, 3,* 11–13.
5. Babkin, V. A., Fedunov, R. G., Ostrouhov, A. A., & Kudryashov, A. V. (2010). Estimation of acid force of molecule 6-methilperhydrotetraline. *Quantum Chemical Calculation of Unique Molecular System.* II. Publisher SUc Volgograd, 74–77.
6. Babkin, V. A., Fedunov, R. G., Ostrouhov, A. A., & Reshetnikov, R. A. (2010). Estimation of acid force of Molecule 7-methilperhydrotetraline, *Quantum Chemical Calculation of Unique Molecular System.* II. Publisher VolSU c. Volgograd, 77–80.
7. Babkin, V. A., Fedunov, R. G., & Ostrouhov, A. A. (2010). About Geometrical and Electronic structure of Molecule Gopan, *Quantum Chemical Calculation of Unique Molecular System.* II. Publisher VolSU c. Volgograd, 80–83.
8. Babkin, V. A., Fedunov, R. G., & Ostrouhov, A. A. (2010). About geometrical and electronic structure of molecule diagopan *Quantum Chemical Calculation of Unique Molecular System.* II. Publisher VolSU c. Volgograd, 83–88.
9. Babkin, V. A., Dmitriev, V. U., & Zaikov, G. E. (2010). Quantum Chemical Calculation of molecule DDT, *Quantum Chemical Calculation of Unique Molecular System. II.* Publisher VolSU c. Volgograd, 13–16.
10. Babkin, V. A., & Andreev, D. S. (2010). Quantum Chemical Calculation of molecule 3-methylcyclopentene by method AB INITIO, *Quantum Chemical Calculation of Unique Molecular System II.* Publisher VolSU c. Volgograd, 101–103.
11. Babkin, V. A., & Andreev, D. S. (2010). Quantum Chemical Calculation of Molecule Cyclopentene by Method AB INITIO, *Quantum Chemical Calculation of Unique Molecular System. II.* Publisher VolSU c. Volgograd, 103–105.

12. Babkin, V. A., & Andreev, D. S. (2010). Quantum Chemical Calculation of Molecule Metylencyclobutane by Method AB INITIO, *Quantum Chemical Calculation of Unique Molecular System. II.* Publisher VolSU, c. Volgograd, 105–107.

13. Babkin, V. A., & Andreev, D. S. (2010). Quantum Chemical Calculation of Molecule 1, 2 -dicyclopropylethylene by method AB INITIO, *Quantum Chemical Calculation of Unique Molecular System. II.* Publisher VolSU c. Volgograd, 108–110.

14. Babkin, V. A., & Andreev, D. S. (2010). Quantum Chemical Calculation of Molecule Iso-Propenylcyclopropane by method AB INITIO, *Quantum Chemical Calculation of Unique Molecular System. II.* Publisher VolSU c. Volgograd, 110–112.

15. Babkin, V. A., Dmitriev, V. Yu., & Zaikov, G. E. (2010). Quantum Chemical Calculation of Molecule Heterolytic Base Uracyl, *Quantum Chemical Calculation of Unique Molecular System. I.* Publisher VolSU c Volgograd, 43–45.

16. Babkin, V. A., Dmitriev, V. Yu., & Zaikov, G. E. (2010). Quantum Chemical Calculation of Molecule Heterolytic Base Adenin, *Quantum Chemical Calculation of Unique Molecular System. I.* Publisher VolSU c. Volgograd, 45–47.

17. Babkin, V. A., Dmitriev, V. Yu., & Zaikov, G. E. (2010). Quantum Chemical Calculation of Molecule Heterolytic Base Guanin, *Quantum Chemical Calculation of Unique Molecular System. I.* Publisher VolSU, c. Volgograd, 47–49.

18. Babkin, V. A., Dmitriev, V. Yu., & Zaikov, G. E. (2010). Quantum Chemical Calculation of Molecule Heterolytic Base Timin, *Quantum Chemical Calculation of Unique Molecular System. I.* Publisher VolSU c. Volgograd, 49–51.

19. Babkin, V. A., Dmitriev, V. Yu., & Zaikov, G. E. (2010). Quantum Chemical Calculation of Molecule Heterolytic Base Cytozin, *Quantum Chemical Calculation of Unique Molecular System. I.* Publisher VolSU c. Volgograd, 51–53.

20. Babkin, V. A., & Andreev, D. S. (2010). Quantum Chemical Calculation of Molecule Isobutilene by Method AB INITIO, *Quantum Chemical Calculation of Unique Molecular System. I.* Publisher VolSU c. Volgograd, 157–159.

21. Babkin, V. A., & Andreev, D. S. (2010). Quantum Chemical Calculation of Molecule 2-methylenbutene-1 by Method AB INITIO, *Quantum Chemical Calculation of Unique Molecular System. I.* Publisher VolSU c. Volgograd, 159–161.

22. Babkin, V. A., & Andreev, D. S. (2010). Quantum Chemical Calculation of Molecule 2-methylbutene-2 by Method AB INITIO, Quantum Chemical Calculation of Unique Molecular System. I. Publisher VolSU c. Volgograd, 161–162.

23. Babkin, V. A., & Andreev, D. S. (2010). Quantum Chemical Calculation of Molecule 2-methylpentene-1 by Method AB INITIO, *Quantum Chemical Calculation of Unique Molecular System. I.* Publisher VolSU c. Volgograd, 162–164.

24. Babkin, V. A., & Andreev, D. S. (2010). Quantum Chemical Calculation of Molecule 2-ethylbutene-1 by Method AB INITIO, *Quantum Chemical Calculation of Unique Molecular System. I.* Publisher VolSU c. Volgograd, 164–166.

CHAPTER 2

QUANTUM-CHEMICAL LECTURE NOTE ON THE MECHANISM OF SYNTHESIS OF 1-[2-(O-ACETYLMETHYL)-3-O-ACETYL-2-ETHYL] – METHYLDICHLORINEPHOSPHITE

V. A. BABKIN, V. U. DMITRIEV, G. A. SAVIN, E. S. TITOVA, and G. E. ZAIKOV

CONTENTS

ABSTRACT

Quantum-chemical research of the mechanism of synthesis of 1-[2-(o-acetylmethyl)-3-o-acetyl-2-ethyl]-methyl dichlorine phosphate for the first time is executed by classical method MNDO. This compound is received by reaction of bimolecular nucleophilic substitution $S_N 2$, proceeding between acetyl chloride and 5-acetyloxy-methyl-2-chlorine-5-ethyl-1,2,3-dioxaphosphorynan. Reaction is endothermic and has barrier character. The size of a power barrier makes 176 kDg/mol.

2.1 AIMS AND BACKGROUNDS

The Mechanism of reaction acylation of bicycle phosphites by chlorine anhydride of carboxylic acids consists of three stages. The mechanism of the first stage is studied by us in Ref. [1]. Results of research of the second stage of this reaction are presented in the yielded work. The second stage represents interaction of acetyl chloride and 5-acetyloxymethyl-2-chlorine-5-ethyl-1,2,3-dioxaphosphorynan in a gas phase:

Now at an electronic level the mechanism of reaction of synthesis of 1-[2-(o-acetyl methyl)-3-o-acetyl-2-ethyl]-methyldichlorinephosphite is not studied. In this connection the aim of this chapter is quantum-chemical research of the mechanism of synthesis of this compound by quantum-chemical method MNDO.

2.2 EXPERIMENTAL PART

The Quantum-chemical semi empirical method MNDO with optimization of geometry on all parameters by the standard gradient method which has been built in PC GAMESS [2] has been chosen for research of the mechanism of synthesis of 1-[2-(o-acetylmethyl)-3-o-acetyl-2-ethyl]-methyl dichlorinephosphite. This method well enough reproduces power characteristics and stability of chemical compounds, including the substances containing multiple bonds [3]. Calculations were carried out in approach to the isolated molecule to a gas phase. The Program MacMolPlt was used for visual representation initial, intermediate and final models [4].

The Mechanism of synthesis of studied compound was investigated by method MNDO. Initial models of components of synthesis 5-acetyloxymethyl-2-chlorine-5-ethyl-1,2,3-dioxaphosphorynan and acetyl chloride settled down on distance 2.8–3.0Å from each other. Any interactions between compounds practically are absent on such distance. Distance R_{O9C29} has been chosen as coordinate of reaction

(Fig. 2.1). This coordinate is most energetically a favorable direction of interaction of initial components. Further, optimization on all parameters of initial components at R_{O9C29} = 2.8Å was carried out. After optimization of value of lengths of bonds and valent corners, values E_0 (the general energy of system) and q_H-charges on atoms along coordinate of reaction R_{O9C29} were fixed and brought in Tables 2.1–2.3. The coordinate of reaction changed from 2.8Å up to 1.3Å at each step of optimization. The step on coordinate of reaction R_{O9C29} has made 0.2Å [5].

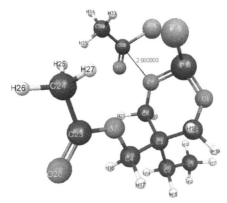

FIGURE 2.1   Initial model of interaction of 5-acetyloxymethyl-2-chlorine-5-ethyl-1,2,3-dioxaphosphorynan and acetyl chloride R_{O9C29} = 2.8Å.

TABLE 2.1   Change of Energy along Coordinate of Reaction R_{O9C29}

R_{O9C29}, Å.	E_0, kDg/mol
2.8	−385.555
2.6	−385.540
2.4	−385.511
2.2	−385.474
2.0	−385.427
1.8	−385.380
1.6	−385.450
1.4	−385.553
1.3	−385.545

TABLE 2.2   Change of Lengths of Bonds along Coordinate of Reaction R_{O9C29}

R_{C29-O9}, Å	R_{C1-C2}	R_{C2-C3}	R_{C3-C4}	R_{C3-C5}	R_{C3-C6}	R_{C4-O7}	R_{C5-O8}	R_{C6-O9}	R_{P10-O8}
2.8	1.53	1.57	1.58	1.58	1.58	1.40	1.39	1.39	1.58
2.6	1.53	1.57	1.58	1.58	1.58	1.40	1.39	1.39	1.58
2.4	1.53	1.57	1.58	1.57	1.58	1.40	1.39	1.40	1.58
2.2	1.53	1.57	1.58	1.57	1.58	1.40	1.39	1.40	1.58
2.0	1.53	1.57	1.58	1.57	1.58	1.40	1.39	1.40	1.57
1.8	1.53	1.57	1.58	1.57	1.58	1.40	1.38	1.42	1.57
1.6	1.53	1.57	1.58	1.58	1.58	1.40	1.38	1.41	1.58
1.4	1.53	1.57	1.59	1.57	1.57	1.40	1.39	1.40	1.57
1.3	1.53	1.57	1.59	1.57	1.57	1.40	1.39	1.41	1.57

R_{C29-O9}, Å	R_{H18-C5}	R_{H19-C5}	$R_{H;20-C6}$	R_{H21-C6}	$R_{C22-P10}$	R_{C23-O7}	$R_{C24-C23}$	$R_{H25-C24}$	$R_{H26-C24}$
2.8	1.12	1.11	1.12	1.12	2.04	1.36	1.52	1.10	1.10
2.6	1.12	1.11	1.12	1.12	2.04	1.36	1.52	1.10	1.10
2.4	1.12	1.11	1.12	1.12	2.04	1.36	1.52	1.10	1.10
2.2	1.12	1.11	1.12	1.12	2.04	1.36	1.52	1.10	1.10
2.0	1.12	1.11	1.12	1.12	2.03	1.36	1.52	1.10	1.10
1.8	1.12	1.11	1.12	1.11	2.03	1.36	1.52	1.10	1.10
1.6	1.12	1.12	1.12	1.12	2.04	1.36	1.52	1.10	1.10
1.4	1.12	1.12	1.12	1.12	2.01	1.36	1.52	1.10	1.10
1.3	1.12	1.12	1.11	1.12	2.01	1.36	1.52	1.10	1.10

R_{C29-O9}, Å	$R_{H27-C24}$	$R_{O28-C23}$	$R_{C30-C29}$	$R_{O31-C29}$	$R_{Cl32-C29}$	$R_{Cl32-P10}$	$R_{H33-C30}$	$R_{H34-C30}$	$R_{H35-C30}$
2.8	1.10	1.22	1.51	1.20	**1.80**	**3.54**	1.10	1.11	1.10
2.6	1.10	1.22	1.51	1.20	**1.81**	**3.34**	1.10	1.11	1.10
2.4	1.10	1.22	1.51	1.20	**1.81**	**3.16**	1.10	1.11	1.10
2.2	1.10	1.22	1.52	1.20	**1.82**	**2.99**	1.10	1.11	1.10
2.0	1.10	1.22	1.52	1.21	**1.84**	**2.83**	1.10	1.11	1.10
1.8	1.10	1.22	1.53	1.21	**1.88**	**2.67**	1.10	1.11	1.10
1.6	1.10	1.22	1.50	1.20	**4.14**	**2.03**	1.10	1.10	1.11
1.4	1.10	1.22	1.52	1.22	**5.76**	**2.00**	1.10	1.10	1.10
1.3	1.10	1.22	1.53	1.23	**5.75**	**2.00**	1.10	1.10	1.10

TABLE 2.3 Change of Charges of System along Coordinate of Reaction R_{O9C29}

$R_{O9C29,}$ Å	2.8	2.6	2.4	2.2	2.0	1.8	1.6	1.4	1.3
C1	0.02	0.02	0.02	0.02	0.02	0.02	0.03	0.03	0.03
C2	0.00	0.00	0.00	0.00	0.00	0.00	0.00	−0.01	−0.01
C3	−0.16	−0.16	−0.16	−0.16	−0.16	−0.15	−0.13	−0.09	−0.09
C4	0.21	0.21	0.20	0.21	0.20	0.20	0.20	0.20	0.20
C5	0.21	0.21	0.21	0.21	0.21	0.22	0.23	0.25	0.25
C6	0.22	0.22	0.22	0.21	0.21	0.20	0.19	0.21	0.22
O7	−0.35	−0.35	−0.35	−0.35	−0.35	−0.35	−0.35	−0.34	−0.34
O8	−0.51	−0.51	−0.51	−0.51	−0.51	−0.52	−0.50	−0.50	−0.51
O9	−0.53	−0.52	−0.53	−0.53	−0.52	−0.50	−0.43	−0.36	−0.32
P10	1.01	1.01	1.01	1.02	1.03	1.07	1.03	0.91	0.91
H11	0.00	0.00	0.00	0.00	0.00	0.00	0.01	0.01	0.01
H12	0.00	0.00	0.00	0.00	0.00	0.00	0.00	0.00	0.00
H13	0.00	0.00	0.00	0.00	0.00	0.00	0.01	0.01	0.01
H14	0.01	0.01	0.01	0.01	0.01	0.01	0.01	0.01	0.01
H15	0.01	0.01	0.01	0.01	0.01	0.01	0.01	0.01	0.01
H16	0.01	0.01	0.01	0.01	0.01	0.01	0.01	0.01	0.01
H17	0.01	0.01	0.01	0.01	0.01	0.02	0.02	0.01	0.01
H18	0.02	0.01	0.02	0.02	0.02	0.02	0.01	0.01	0.01
H19	0.03	0.03	0.04	0.04	0.04	0.04	0.03	0.01	0.01
H20	0.01	0.01	0.02	0.02	0.02	0.03	0.02	0.01	0.02
H21	0.04	0.03	0.03	0.03	0.03	0.04	0.01	0.01	0.02
Cl22	−0.42	−0.41	−0.41	−0.41	−0.41	−0.40	−0.48	−0.36	−0.36
C23	0.35	0.35	0.35	0.35	0.35	0.35	0.35	0.35	0.35
C24	0.05	0.05	0.05	0.05	0.05	0.05	0.05	0.05	0.05
H25	0.03	0.03	0.03	0.03	0.03	0.03	0.02	0.03	0.03
H26	0.03	0.03	0.03	0.03	0.03	0.03	0.03	0.03	0.03
H27	0.04	0.04	0.04	0.04	0.04	0.04	0.03	0.03	0.03
O28	−0.35	−0.35	−0.35	−0.35	−0.35	−0.35	−0.35	−0.35	−0.35
C29	0.34	0.34	0.35	0.37	0.39	0.42	0.38	0.35	0.35
C30	0.02	0.02	0.03	0.03	0.03	0.03	0.05	0.05	0.05
O31	−0.23	−0.24	−0.24	−0.25	−0.28	−0.32	−0.21	−0.33	−0.38
Cl32	−0.24	−0.24	−0.25	−0.27	−0.29	−0.34	−0.42	−0.31	−0.31
H33	0.05	0.05	0.05	0.05	0.04	0.04	0.05	0.03	0.02
H34	0.03	0.03	0.03	0.03	0.03	0.04	0.06	0.03	0.03
H35	0.03	0.03	0.03	0.03	0.02	0.02	0.04	0.03	0.02

TABLE 2.4   Change of Valence Corners along Coordinate of Reaction R_{O9C29}

Coordinate of reaction. R_{O9C29}, Å.	2.8	2.6	2.4	2.2	2.0	1.8	1.6	1.4	1.3
C(3)C(2)C(1)	119	120	120	120	120	120	120	120	120
C(4)C(3)C(2)	105	105	104	104	104	104	104	104	104
C(5)C(3)C(2)	110	110	110	110	110	110	110	111	111
C(6)C(3)C(2)	108	108	108	108	108	108	108	108	108
O(7)C(4)C(3)	110	110	110	110	111	111	111	111	111
O(8)C(5)C(3)	113	113	113	112	112	112	113	112	111
O(9)C(6)C(3)	115	116	117	117	117	117	113	111	111
P(10)O(8)C(5)	127	128	128	128	128	129	125	129	129
H(11)C(1)C(2)	112	112	112	112	112	112	112	112	112
H(12)C(1)H(11)	107	107	107	107	107	107	107	106	106
H(13)C(1)C(12)	107	107	107	107	106	106	107	106	106
H(14)C(2)C(1)	107	107	107	107	107	107	107	106	106
H(15)C(2)H(14)	105	105	105	105	105	105	105	105	105
H(16)C(4)O(7)	110	110	110	110	110	110	109	110	109
H(17)C(4)H(16)	106	106	106	106	106	106	106	106	106
H(18)C(5)O(8)	109	109	109	109	109	109	109	109	109
H(19)C(5)H(18)	104	104	104	104	104	104	105	106	106
H(20)C(6)O(9)	107	107	107	107	107	107	108	109	108
H(21)C(6)H(20)	105	105	105	105	106	106	105	106	106
Cl(22)P(10)O(8)	105	105	104	104	104	104	103	105	105
C(23)O(7)C(4)	125	124	124	124	124	124	124	124	124
C(24)C(23)O(7)	113	113	113	113	113	113	113	113	113
H(25)C(24)C(23)	110	110	110	110	110	111	110	110	110
H(26)C(24)H(25)	108	108	108	108	108	108	108	108	108
H(27)C(24)H(25)	108	108	108	108	108	108	108	108	108

TABLE 2.4 *(Continued)*

Coordinate of reaction. R_{O9C29}, Å.	2.8	2.6	2.4	2.2	2.0	1.8	1.6	1.4	1.3
O(28)C(23)O(7)	119	119	119	119	119	119	119	120	120
C(29)O(9)C(6)	115	116	117	118	118	118	118	124	126
C(30)C(29)O(9)	97	97	97	98	100	104	112	112	114
O(31)C(29)O(30)	128	128	128	128	127	126	135	127	123
Cl(32)C(29)O(9)	93	93	94	94	95	95	54	53	53
H(33)C(30)C(29)	111	112	112	112	113	113	112	111	109
H(34)C(30)H(32)	97	99	97	97	95	93	60	41	84
H(35)C(30)H(34)	108	109	108	108	108	108	108	108	107

2.3 RESULTS OF CALCULATIONS

The Optimized geometrical and electronic structure of initial models (a stage of co-ordination), an intermediate condition of system (during the moment of separation atom Cl_{32}) and a final condition of molecule of 1-[2-(o-acetyl methyl)-3-o-acetyl-2-ethyl]-methyldichlorinephosphite, are presented on Figs. 2.1–2.3. Changes of lengths of bonds, valent corners, on atoms along coordinate of reaction R_{O9C29} with step 0.2Å are shown to the general energy of all molecular system and charges in Tables 2.1–2.3 and on schedules (Figs. 2.4–2.8). Essential changes to system of initial components at steps 1–6 (R_{O9C29} changes from 2.8Å up to 1.8Å) does not occur. Their mutual orientation from each other goes at this stage. We define this stage as a stage of coordination. Simultaneous break of bond C_{29}–Cl_{32} and almost full break of bond P_{10}–O_9 occurs at II stage of interaction (6–7 steps, R_{O9C29} change from 1.8Å up to 1.4Å). Lengths of these bonds change from 1.88Å up to 4.14Å and from 1.64Å up to 1, 93Å accordingly (Table 2.2). We define this stage, as a stage of break of bonds. Atom P_{10} of 5-acetyloxymethyl-2-chlorine-5-ethyl-1,2,3-dioxaphosphorynan is attacked by atom Cl_{32} of acetyl chloride at a stage III. Covalent bonds P_{10}–Cl_{32} (2.00Å) and O_9–C_{29} (R_{O9C29} = 1.4Å) are formed. This stage is final. Formation of 1–[2–(o-acetylmethyl)-3-o-acetyl-2-ethyl]-methyldichlorinephosphite occurs.

Change of charges on atoms directly participating in reaction P_{10}, C_{29} and Cl_{32}, and also change E_0 along coordinate of reaction R_{O9C29} is presented on Figs. 2.4–2.8 and in Table 2.1. Reaction has barrier character (Fig 2.4). The size of a barrier makes 176 kDg/mol. Charges on atoms P_{10} and C_{29} change according to change E_0 (Figs 2.5 and 2.7). Charges reach the maximal values during the moment of break of bonds

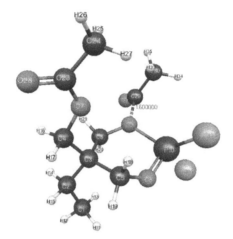

FIGURE 2.2   The Model of a stage of break of bonds (a transition state) $R_{O9C29} = 1.6\text{Å}$.

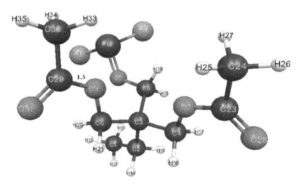

FIGURE 2.3   The Model of formation of 1-[2-(o-acetylmethyl)-3-o-acetyl-2-ethyl]-methyldichlorinephosphite. $R_{O9C29} = 1.3\text{Å}$.

C_{29}–Cl_{32} and P_{10}–O_9. The Negative charge on atom Cl_{32} is inversely to change E_0. It reaches the maximal value (on the module) during the moment of break of the same bonds. We analyze behavior of atoms directly participating in reaction of synthesis of 1-[2-(o-acetylmethyl)-3-o-acetyl-2-ethyl]-methyldichlorinephosphite P_{10}, O_9, C_{29} and Cl_{32}, change of charges on these atoms and power of reaction. We do a conclusion, that the mechanism of studied synthesis represents the coordinated process with simultaneous break of bonds P_{10}–O_9 and C_{29}–Cl_{32} and formation of new bonds P_{10}–Cl_{32} and C_{29}–O_9. It is a process of nucleophylic substitution S_N2. It is similar to the mechanism of synthesis of the first stage of acidation of bicyclophosphites by chlorine anhydrides of carboxylic acids, to the mechanism of synthesis of 5-acety-loxymethyl-2-chlorine-5-ethyl-1,2,3-dioxaphosphorynan.

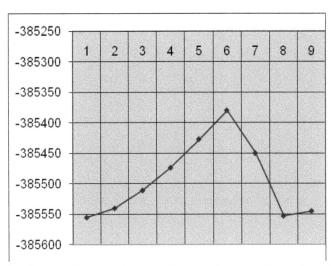

FIGURE 2.4   Change of the general energy of system along coordinate of reaction R_{O9C29}.

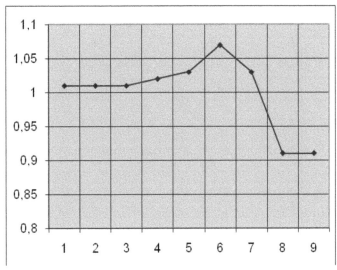

FIGURE 2.5   Change of a positive charge on atom of phosphorus P_{10} along coordinate of reaction R_{O9C29}.

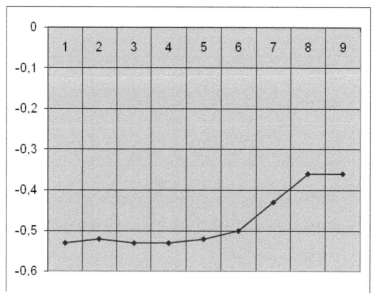

FIGURE 2.6   Change of a negative charge on atom O_9 along coordinate of reaction R_{O9C29}.

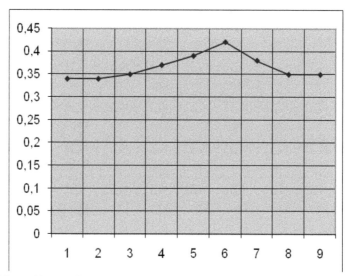

FIGURE 2.7   Change of a positive charge on atom of carbon C_{29} along coordinate of reaction R_{O9C29}.

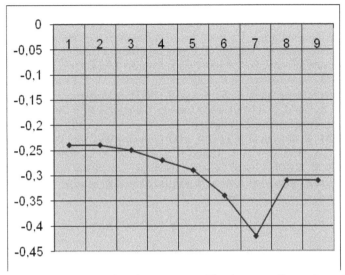

FIGURE 2.8   Change of a negative charge on atom Cl_{32} along coordinate of reaction R_{O9C29}.

Thus, the mechanism of synthesis of the second stage of acidation of bicyclo-phosphites for the first time is studied by quantum-chemical semi empirical method MNDO. It is shown, that synthesis of this compound result of the coordinated interactions of acetyl chloride and 5-acetyloxymethyl-2-chlorine-5-ethyl-1,2,3-dioxa-phosphorynan on the mechanism of nucleophylic substitution SN2. It is positioned, that this reaction is endothermic and has barrier character. It will qualitatively be coordinated with experiment. The size of an energy barrier of studied reaction is equal 176 kDg/mol.

KEYWORDS

- 1-[2-(o-acetylmethyl)-3-o-acetyl-2-ethyl]-methyldichlorinephosphite
- 5-acetyloxymethyl-2-chlorine-5-ethyl-1,2,3-dioxaphosphorynan
- Acetyl chloride
- Method MNDO
- Quantum-chemical research
- The mechanism of synthesis

REFERENCES

1. Babkin, V. A., Dmitriev, V. U., Savin, G. A., & Zaikov, G. E. (2009). Estimation of acid force of components of synthesis of 5-acetyloxymethyl-2-chlorine-5-ethyl-1,2,3-dioxaphosphorynan Moscow, *Encyclopedia of the Engineer Chemist, 13,* 11–13

2. Schmidt, M. W., Baldrosge, K. K., Elbert, J. A., Gordon, M. S., Enseh, J. H., Koseki, S., Matsvnaga, N., Nguyen, K. A., Su, S. J. et al. (1993). *J Computer Chem. 14,* 1347–1363.

3. Clark, T., & World, M. (1990). *The Computer Chemistry*, 383p Quantum-chemical research of the mechanism of synthesis.

4. Bode, B. M., & Gordon, M. S. (1998). *J Mol Graphics Mod, 16,* 133–138.

5. Babkin, V. A., Rachimov, A. I., Titova, E. S., Fedunov, R. G., Reshetnikov, R. A., Belousova, V. S., & Zaikov, G. E. Quontum-chemical researches of the mechanism of synthesis 2-methil (benzyl)-tio-4-methil(benzil)oxipyrimidine Izhevsk. *The Chemical Physics and Mesoscopy, 9,* 263–276.

CHAPTER 3

A LECTURE NOTE ON THE EFFECT OF ANTIOXIDANT COMPOUNDS ON OXIDATIVE STRESS IN UNICELLULAR AQUATIC ORGANISMS

O. V. KARPUKHINA, K. Z. GUMARGALIEVA,
and A. N. INOZEMTSEV

CONTENTS

ABSTRACT

Toxic effect of heavy metal salts and hydrogen peroxide lipid per oxidation inducers was studied on the culture of *Paramecium caudatum* cells. A significant protective effect of substances with antioxidant effect (ascorbic acid, piracetam) in the experimental model of induced oxidative stress in unicellular infusoria is established.

3.1   INTRODUCTION

Heavy metals are potentially dangerous toxicants which are the destabilizing factor in the ecological system of established biocoenosis. lead, cadmium, mercury, cobalt, and other heavy metals may be inducers of the oxidative stress, which is based on formation of excessive quantity of free radicals [1, 2]. Their reactivity is extremely high and initiates chain oxidation reaction. Free radicals become the reason for serious functional disorders, because various cell components are damaged [3]. Initiation of lipid peroxidation in biological membranes is an example and promotes disturbance of their structure and penetrability increase. Specialized enzymatic and nonenzymatic antioxidant systems are protection against free radicals [4].

Determination of substances which action is aimed at normalizing metabolic processes, blockage of pathological free-radical processes is the important branch of investigations of organism adaptation to the impact of toxic agents. It is common knowledge that such compound is ascorbic acid able to react with superoxide and hydroxyl radicals and thus decreases their concentration in the cell, preventing development of the oxidant stress [5]. In tests on rats, it was found that ascorbic acid caused protective antioxidant effect on formation of conditioned responses under toxic effect of joint action of heavy metal salts (Co, Pb, and Cd) and neurotropic preparation piracetam [6, 8]. It has been found that catalytic action of heavy metal salts affects piracetam structural characteristics and, consequently, distorts its curative effect [7]. It is shown in vitro that ascorbic acid prevents heavy metal salts interaction with the drug preparation structure, and injection of the antioxidant into animals has eliminated the negative effect of heavy metals on formation of adaptive reactions [8].

The complex organism responses which *is protective and* provides adaptation to varying conditions in response to the impact of various adverse factors, including toxic agents, is based on global fundamental mechanisms. Unicellular infusoria may be considered as simple receptor-effectors systems capable of rapidly respond to a chemical action by entire set of biological, physiological and biochemical changes: chemotaxis, backward mutation of the ciliate activity, the rate of reproduction, and phagocytic activity. The aim of this chapter was to study the effect of toxic agents the inductors of oxidative stress and protective action of antioxidant agents, on the vital activity of unicellular aquatic organisms *Paramecium caudatum.*

3.2 EXPERIMENTAL PART

Paramecium caudatum (the wild strain) cell culture was cultivated in the Lozina-Lozinsky medium with addition of nutritive medium yeasts containing *Saccharomyces cerevisiae* yeasts. The oxidative stress was induced by: cadmium chloride, lead acetate, and hydrogen peroxide (H_2O_2). Infusorians sensitivity to toxic agents was determined by time of their death established by protozoa motion cessation, often accompanied by cell deformation and cytolysis. The exposure time was 2 hours. The control in all tests was the number of cells in 10 ml of the medium containing the intact culture of infusorian (without oxidative stress induction).

Total number of cells in 10 ml of the medium containing infusorian was determined in the Goryaev counting chamber. Cells sampled at the stationary phase of growth were incubated in the medium with the chemical substance added at (20 ± 2) °C temperature during 15 days (chronic effect).

The effect of toxic agents was studied in several concentrations (0.05, 0.025, 0.005, 0.0025, and 0.00125 mg/L); the effect of substances with antioxidant properties (ascorbic acid, piracetam) was studied at concentrations of 1, 10, 50 µM, and 1 mM.

3.3 RESULTS AND DISCUSSION

As established in the experiments performed, in the initial 30 min of incubation, cadmium chloride and lead acetate (0.05 mg/L) caused immediate motion cessation of infusorian with the shape change to rod that showed structural changes in the cell membranes. Hydrogen peroxide (0.05 mg/L) initiated degradation of the lipid part of the unicellular organism membrane: gross morphological changes in the cellular membranes of infusorian as multiple sub globose projections (the cell membrane "vacuolization") occurred, which then broke and caused death of the cell-organism. The number of species in control does not change during the entire test (Fig. 3.1). Hydrogen peroxide injection to the medium with infusorians caused 3-fold decrease of the number of species in the first minute of the test. Decrease in population of test cells indicates the destructive membrane pathology in which occurrence free-radical oxidation processes are of importance. By 60th minute of the test, 100% infusorian death was observed. The toxic effect of heavy metal salts was less expressed.

The goal of the following test was the study of hydrogen peroxide and heavy metal salts effect on reproduction of Paramecium cells. Since the cell culture death at 0.05 and 0.025 mg/L concentration was observed (Fig. 3.1), concentrations of toxic agents in tests on the cell reproduction were 20-fold decreased. At this concentration, a considerable part of cells survived during the first 24 hours of the test (Fig. 3.2). For the cell incubation during 15 days, media with hydrogen peroxide and cadmium chloride, in which infusorian have not reproduced even the first generation, were found the most toxic ones (Fig. 3.2).

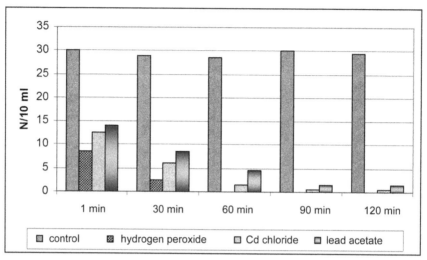

FIGURE 3.1 Toxic effect of chemical agents on *Paramecium caudatum* cell culture *(abscissa axis test duration, axis of ordinates the number of species in 10 mL of the test medium).*

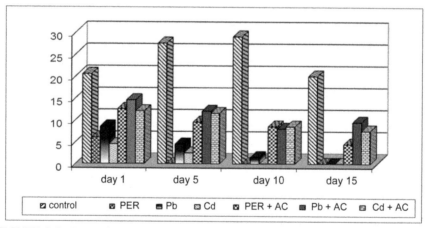

FIGURE 3.2 Dynamics of *Paramecium caudatum* population changes in the medium containing various chemical agents: control-normal saline; PER-hydrogen peroxide; Pb-lead diacetate; AC – ascorbic acid.

By the 5th day, infusorians incubated in hydrogen peroxide died. Adding of ascorbic acid (1 μM) to the cell medium subjected to hydrogen peroxide action almost two-fold increased the number of species in thc first 24 hours. At the same time, ascorbic acid prevented death of culture cells that was noted in all test days (Figs. 3.2 and 3.3).

In the medium with a heavy metal salt added, 100% death rate of the culture was observed on the 15th day.

By the 15th day, only an insignificant decrease in the species population was observed, when ascorbic acid or lead acetate was added to the cell culture. Thus, here ascorbic acid manifested similar antioxidant activity as against hydrogen peroxide.

To prove efficiency of an antioxidant agent with the action mechanism different from ascorbic acid, we have studied Paramecium caudatum in the medium with added piracetam the compound with nonspecific antioxidant activity [9] (Fig. 3.3).

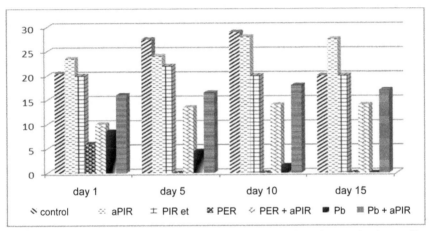

FIGURE 3.3 Dynamics of *Paramecium caudatum* population changes in the medium containing various chemical agents: control normal saline; PER–hydrogen peroxide; Pb–lead diacetate; AC-ascorbic acid; PIR et piracetam solution (drag form); a PIR – structurally modified piracetam solution.

In these tests, the efficiency of the piracetam drug form (PIR et) and it's structurally modified form (a PIR). Both substances prevented death of infusorian cells due to hydrogen peroxide and heavy metal salt action. The antioxidant effect of piracetam was less expressed as compared with protective action of ascorbic acid. Nevertheless, by the 15th day survivability of species with injection of a PIR into the medium with infusorian reached 100%.

3.4 CONCLUSION

The experiments performed have shown that responses of unicellular aquatic organisms are effective for obtaining primary information on cytotoxicity of substances (heavy metal salts, peroxides, etc.).

Ascorbic acid and membrane-stabilizing antioxidant substance piracetam significantly increase cell nonresponsiveness to heavy metals that proves the important role of antioxidant compounds in free-radical oxidation processes in the organism cell.

KEYWORDS

- **Antioxidants**
- **Heavy Metals**
- **Oxidative Stress**
- **Paramecium Caudatum**

REFERENCES

1. Simmons, S. O., Fan, C. Y., & Ramabhadran, R. (2009). Cellular Stress Response Pathway System as a Sentinel Ensemble in toxicological screening, *Toxically Sci., 111(2)*, 202.
2. Leonard, S. S., Harris, G. K., & Shi, X. L. (2004). Metal induced oxidative stress and signal transduction, *Free Rad Boil Med., 37(12)*, 1921.
3. Pryor, W. A. (1986). Oxy-radicals and related species their formation lifetimes and reactions. *Annual Review of Physiology, 48*, 657.
4. Flora, S. J. (2009). Structural chemical and biological aspects of antioxidants for strategies against metal and metalloid exposure, *Oxid Med Cell, 2(4)*, 191.
5. Davies, M., Austin, J., & Partridge, D. (1991). Vitamin C. Its Chemistry and Biochemistry Royal Society of Chemistry London UK.
6. Bokieva, S. B., Karpukhina, O. V., Gumargalieva, K. Z., & Inozemtsev, A. N. (2012). Combined effects of heavy metals and Piracetam destroying the adaptive behavior formations. Proceedings of Gorsky State university of Agriculture, *4(4)*, 194.
7. Karpukhina, O. V., Gumargalieva, K. Z., Soloviev, A. G., inozemtsev, A. N. (2004). Effects of Lead Diacetate on Structure Transformation and Functional Properties of Piracetam, *J. Environmental Protection Ecology, 5(3)*, 577.
8. Karpukhina, O. V., Gumargalieva, K. Z., bokieva, S. B., Kalyuzhny, A. L., & Inozemtsev, A. N. (2011). Ascorbic acid protects the body from toxic effects of cadmium. *"High Technology Basic and applied Researches in Physiology and Medicine."* Saint-Petersburg Russia, *3(4)*, 166.
9. Gouliaev, A. H., & Senning, A. (1994). Piracetam and other structurally related nootropics, *Brain Res Rev., 19(2)*, 180.

CHAPTER 4

ON THE DIFFUSION OF AMMONIUM SALTS IN AQUEOUS SOLUTIONS

LUÍS M. P. VERÍSSIMO, ABÍLIO J. F. N. SOBRAL,
ANA C. F. RIBEIRO, CARMEN TEIJEIRO, and M. A. ESTESO

CONTENTS

ABSTRACT

Diffusion coefficients of ammonium salts in aqueous solutions are theoretically estimated from the Onsager-Fuoss model. The influence of the ion size parameter a (mean distance of closest approach of ions), as well as of both the thermodynamic and the mobility factors on the variation of diffusion coefficients with concentration, is discussed. The aim of this chapter is to contribute to a better knowledge of the structure of these systems.

4.1 INTRODUCTION

Aqueous solutions of ammonium compounds have become important systems in fundamentals and applied research due to their wide range of applications related to food processing (e.g., [1, 2]), human and veterinary medicine and pharmaceutical chemistry (e.g., [3, 4]). However, despite the many reasons justifying the importance of these salts, the understanding of these complex systems has not yet been well established. Therefore, their characterization is very important as it helps to better understand their structure and to model them for practical applications.

We are particularly interested in the study of the diffusion processes in electrolyte solutions, because it is important for fundamental reasons, helping to understand the nature of aqueous electrolyte structure and the behavior of electrolytes in solution, as well as supplying the scientific and technological communities with data on these important parameters in the solution transport processes [5–7]. In fact, the scarcity of diffusion coefficients in the scientific literature, arising from both the difficulty of their accurate experimental measurement and the impracticability of their determination by theoretical procedures, allied to their industrial and research need, well justifies efforts in their theoretical estimation.

By following a careful literature search, only a few experimental data, D, were found for some systems involving ammonium ion (e.g., [8]), and no theory on diffusion in electrolyte solutions is capable of giving generally reliable data relative to the magnitude of the diffusion coefficient. However, for estimating purposes, and as an initial approach to the experimental mutual diffusion coefficients, theoretical values can be estimated by using the Onsager-Fuoss model [9]. Estimations from this theory are adequate for these electrolytes in aqueous dilute solutions ($c \leq 0.005$ mol dm^{-3}), as has been shown for other similar systems, where the theoretical data are consistent with experimental results (deviations, in general, $\leq 3\%$) [10, 11].

This chapter reports theoretical data for differential binary mutual diffusion coefficients of 20 systems containing ammonium ion at different concentrations (from 0.000 to 0.005 mol dm^{-3}), estimated from the Onsager-Fuoss equation, the thermodynamic factor, F_T, and the mobility parameter, F_M. For these estimations, different methods to obtain an adequate value for the mean distance of closest approach of ions parameter, a, were used [10–15].

4.2 ESTIMATIONS OF THE DIFFUSION COEFFICIENTS BY USING THE ONSAGER-FUOSS EQUATION

The estimation of the diffusion coefficients of the ammonium salts in aqueous solutions can be made on the basis of the Onsager-Fuoss model (Eq. (1)) [9], by taking into account that D is a product of both a kinetic F_M (or molar mobility coefficient of a diffusing substance) and a thermodynamic factors, F_T (or gradient of the free energy). Thus, two different effects can control the diffusion process: the ionic mobility and the gradient of free energy.

$$D = F_M \times F_T \tag{1}$$

where

$$F_T = c\frac{\partial \mu}{\partial c} = \left(1 + c\frac{\partial \ln \gamma}{\partial c}\right) \tag{2}$$

and

$$F_M = (D^0 + \Delta_1 + \Delta_2) \tag{3}$$

μ and γ represent the chemical potential and the thermodynamic activity coefficient of the solute, respectively). Δ_1 and Δ_2 represent the first and second-order electrophoretic terms defined by Eq. (4),

$$\Delta n = k_B T A_n \frac{\left(z_1^n t_2^0 + z_2^n t_1^0\right)^2}{|z_1 z_2| a^n} \tag{4}$$

where k_B is the Boltzmann's constant, T is the absolute temperature, A_n are functions of the dielectric constant, of the solvent viscosity, of the temperature, and of the dimensionless concentration-dependent quantity (ka), being k the reciprocal of average radius of the ionic atmosphere and a the mean distance of closest approach of ions; t_1^0 and t_2^0 are the limiting transport numbers of the cation and the anion, respectively.

4.3 DIFFERENT EXPERIMENTAL AND THEORETICAL METHODS OF ESTIMATION OF THE PARAMETER A

Though there is no a direct method for measuring the ion size parameter a (mean distance of closest approach), it may be well estimated from the data of ionic sizes presented by Kielland [16], as the mean value of the effective radii of the hydrated ionic species of the electrolyte (1st column in Table 4.1). The diameters of inorganic ions, hydrated to a different extent, have been calculated by two different methods: (i) from the crystal radius and deformability, accordingly to Bonino's equation for cations [16]; and (ii) from the ionic mobilities [16].

Molecular modeling studies are also important tools to estimate these parameters. They are a valuable tool to interpret atom or ion dynamic relations and they are simpler than *ab initio* calculations but yet they gave very close results. For that reason they are adequate to evaluate dynamic processes like solvation changes and mean distances of approach between species in solution. Among the MM methods [17], MM+ is a reference in the area and was used in this study. The results obtained are summarized in the second and third columns in Table 4.1, presenting the results for vacuum and inside a water box of 216 water molecules, respectively. The fourth column represents the ion distances obtained by *ab initio* (Restricted Hartree–Fock) using a small basis (in vacuum) [18]. Calculations were performed in a HP Z620 workstation using the HyperChemv7.5 software package from Hypercube Inc., 2000, USA. The geometry optimization used a Polak-Ribiere conjugated gradient algorithm for energy minimization in vacuum, with a final gradient of 0.1kcal/Å mol. The ion distances were calculates after geometry minimization for both molecular mechanics and *ab initio* method.

4.4   RESULTS AND DISCUSSION

Table 4.1 summarizes the values of the ion size parameter, *a*, of 20 ammonium salts in aqueous solutions, determined from different experimental techniques and/ or theoretical approaches for every electrolyte.

TABLE 4.1   Summary of the Mean Distance of Closest Approach Values ($a/10^{-10}$ m) for Ammonium Salts in Aqueous Solutions, Estimated from Experimental Data, Ionic Radius and other Theoretical Approaches

Electrolyte	Kielland[a]	Molecular mechanics MM+[b]	Molecular mechanics MM[b]	ab initio[c]
NH_4Br	2.8	3.8	3.7	3.0
NH_4Cl	2.8	4.0	3.7	2.9
NH_4F	3.0	3.7	3.7	2.7
NH_4I	2.8	4.0	4.0	3.3
NH_4CHO_2	3.0	4.4	4.4	2.7
$NH_4C_2H_3O_2$	3.5	4.1	4.0	2.7

TABLE 4.1 *(Continued)*

Electrolyte	Kielland[a]	Molecular mechanics MM+[b]	Molecular mechanics MM[b]	ab initio[c]
$(NH_4)_2CO_3$	3.5	3.3	3.6	2.7
$(NH_4)_2C_2O_4$	3.5	3.4	3.8	2.6
NH_4ClO_4	3.0	4.0	4.1	3.1
NH_4HCO_3	3.4	3.8	3.1	2.6
$NH_4H_2PO_4$	3.4	3.4	3.3	2.6
$(NH_4)_2HPO_4$	3.3	3.4	3.4	2.6
NH_4HSO_3	3.4	3.4	3.5	2.5
NH_4IO_3	3.4	3.5	3.6	2.6
NH_4NO_2	2.8	3.6	3.7	2.7
NH_4NO_3	2.8	3.5	3.6	2.7
NH_4OH	3.0	3.3	3.7	2.9
NH_4SCN	3.0	3.7	3.6	3.2
$(NH_4)_2SO_3$	3.5	4.2	5.8	2.6
$(NH_4)_2SO_4$	3.3	3.2	3.3	2.6

[a] See [16]. [b] Molecular Mechanics MM+ [17]. [c]*RHF small basis set 3–21G in vacuum* [18].

Table 4.1 also shows that, in general, for ammonium salts the values of this parameter obtained by MM+ (in vacuum) and *ab initio* (in vacuum) theoretical methods are approximately equal to those obtained from the Kielland's data. The ion-ion distances obtained by *ab initio* (RHF), using a small basis in vacuum, showed a tendency to be smaller than those obtained by the molecular mechanics methods in aqueous media. Considering that in aqueous solution the ions are generally hydrated, a may be greater than the sum of the crystallographic radii of the ions, and less than the sum of the radii of the hydrated ions; however, from these data, a values are most probably close to the first limit. The smaller differences between them can be explained by their limitations. For example, once Kielland equations involve ionic mobilities (or phenomenological coefficients), which are rigorously valid only at very high dilution. Thus, the ion-ion and hydrodynamic interactions (not considered in this model) may actually influence the phenomenological coefficients and ionic mobilities and, consequently, lead to obtain non-real values of the parameter a.

Despite of these limitations, which make no possible to know a estimated values with accuracy, it is possible to have an idea of their possible range of values, considered all of these methods reasonable compromises. Consequently, for each electrolyte, either the use of a given a value from one specific method of estimation, or the use of an average value from all of the methods is legitimate (Table 4.1). In our case, we have used the average values of that parameter in the estimation of diffusion coefficients (Table 4.2), because almost all D_{OF} values obtained from different a values, are close each other (deviations, in general, $< 2\%$).

Considering Eqs. (1)–(4), we have estimated D_{OF}, F_M and $(\Delta_1 + \Delta_2)$, by using the average values of the a parameter, and F_T from available activity coefficient data [6] (Table 4.2). While F_T decreases as concentration increases, F_M increases for the same concentration range and $(\Delta_1 + \Delta_2)$ values are small.

TABLE 4.2 D_{OF}, F_T, F_M, Δ_c and Δ_a Values for Ammonium Salts, Calculated from Eqs. (1)–(4) at 298.15 K[a]

Electrolyte	c [a]	D_{OF} [b]	F_M [c]	$(\Delta_1 + \Delta_2)$ [c]	F_T [d]
NH$_4$Br	0.000	$(a = 3.3\times10^{-10}$ m$)$	$(a = 3.3\times10^{-10}$ m$)$	$(a = 3.3\times10^{-10}$ m$)$	$(a = 3.3\times10^{-10}$ m$)$ 1.000
	0.001	2.023	2.023	0.000	0.983
	0.002	1.990	2.024	0.001	0.976
	0.003	1.980	2.029	0.006	0.972
	0.004	1.972	2.029	0.006	0.968
	0.005	1.966	2.031	0.008	0.965
		1.960	2.036	0.013	
NH$_4$Cl	0.000	$(a = 2.9\times10^{-10}$ m$)$	1.996	0.000	$(a = 2.9\times10^{-10}$ m$)$ 1.000
	0.001	1.996	1.999	0.003	0.983
	0.002	1.965	2.002	0.006	0.976
	0.003	1.954	2.004	0.008	0.971
	0.004	1.946	2.006	0.010	0.967
	0.005	1.940	2.006	0.010	0.964
		1.934			

TABLE 4.2 *(Continued)*

Electrolyte	c [a]	D_{OF} [b]	F_M [c]	$(\Delta_1 + \Delta_2)$ [c]	F_T [d]
NH_4F	0.000	$(a = 3.3 \times 10^{-10}$ m)	1.684	0.000	$(a = 2.9 \times 10^{-10}$ m) 1.000
	0.001	1.684	1.687	0.003	0.983
	0.002	1.658	1.688	0.004	0.976
	0.003	1.648	1.689	0.005	0.972
	0.004	1.642	1.690	0.006	0.968
	0.005	1.636	1.691	0.007	0.965
		1.632			
NH_4I	0.000	$(a = 3.5 \times 10^{-10}$ m)	2.004	0.000	$(a = 3.5 \times 10^{-10}$ m) 1.000
	0.001	2.004	2.007	0.003	0.983
	0.002	1.973	2.009	0.005	0.976
	0.003	1.961	2.012	0.008	0.971
	0.004	1.954	2.013	0.009	0.967
	0.005	1.947	2.011	0.007	0.965
		1.941			
NH_4CHO_2	0.000	$(a = 3.6 \times 10^{-10}$ m)	1.679	0.000	$(a = 3.6 \times 10^{-10}$ m) 1.000
	0.001	1.679	1.682	0.003	0.983
	0.002	1.653	1.682	0.003	0.977
	0.003	1.643	1.684	0.005	0.972
	0.004	1.637	1.685	0.006	0.968
	0.005	1.631	1.686	0.007	0.965
		1.627			
$NH_4C_2H_3O_2$	0.000	$(a = 3.6 \times 10^{-10}$ m)	1.403	0.000	$(a = 3.6 \times 10^{-10}$ m) 1.000
	0.001	1.403	1.405	0.002	0.983
	0.002	1.381	1.404	0.001	0.977
	0.003	1.372	1.405	0.002	0.972
	0.004	1.366	1.406	0.003	0.968
	0.005	1.361	1.408	0.005	0.965
		1.359			

TABLE 4.2 *(Continued)*

Electrolyte	c [a]	D_{OF} [b]	F_M [c]	$(\Delta_1 + \Delta_2)$ [c]	F_T [d]
$(NH_4)_2CO_3$	0.000	$(a = 3.5 \times 10^{-10}$ m)	1.426	0.000	$(a = 3.6 \times 10^{-10}$ m) 1.000
	0.001	1.426	1.438	0.012	0.943
	0.002	1.356	1.444	0.018	0.924
	0.003	1.334	1.449	0.023	0.910
	0.004	1.319	1.454	0.028	0.899
	0.005	1.307	1.458	0.032	0.890
		1.298			
$(NH_4)_2C_2O_4$	0.000	$(a = 3.5 \times 10^{-10}$ m)	1.172	0.000	$(a = 3.5 \times 10^{-10}$ m) 1.000
	0.001	1.172	1.210	0.038	0.920
	0.002	1.113	1.194	0.022	0.917
	0.003	1.095	1.203	0.031	0.900
	0.004	1.083	1.202	0.030	0.894
	0.005	1.075	1.216	0.044	0.878
		1.068			
NH_4ClO_4	0.000	$(a = 3.5 \times 10^{-10}$ m)	1.873	0.000	$(a = 3.5*10^{-10}$ m) 1.000
	0.001	1.873	1.876	0.003	0.983
	0.002	1.844	1.876	0.003	0.977
	0.003	1.833	1.880	0.007	0.972
	0.004	1.827	1.880	0.007	0.968
	0.005	1.820	1.881	0.008	0.965
		1.815			
NH_4HCO_3	0.000	$(a = 3.2 \times 10^{-10}$ m)	1.477	0.000	$(a = 3.2 \times 10^{-10}$ m) 1.000
	0.001	1.477	1.478	0.001	0.983
	0.002	1.453	1.481	0.004	0.976
	0.003	1.445	1.481	0.004	0.972
	0.004	1.440	1.482	0.005	0.968
	0.005	1.435	1.482	0.005	0.965
		1.430			

TABLE 4.2   *(Continued)*

Electrolyte	c [a]	D_{OF} [b]	F_M [c]	$(\Delta_1 + \Delta_2)$ [c]	F_T [d]
		$(a = 3.2\times10^{-10}$ m)			$(a = 3.2\times10^{-10}$ m)
$NH_4H_2PO_4$	0.000	1.288	1.288	0.000	1.000
	0.001	1.267	1.290	0.002	0.982
	0.002	1.259	1.291	0.003	0.975
	0.003	1.253	1.292	0.004	0.970
	0.004	1.248	1.293	0.005	0.965
	0.005	1.245	1.297	0.009	0.960
		$(a = 3.2\times10^{-10}$ m)			$(a = 3.2\times10^{-10}$ m)
$(NH_4)_2HPO_4$	0.000	1.283	1.283	0.000	1.000
	0.001	1.220	1.294	0.011	0.943
	0.002	1.200	1.300	0.017	0.923
	0.003	1.187	1.307	0.024	0.908
	0.004	1.176	1.311	0.028	0.897
	0.005	1.165	1.312	0.029	0.888
		$(a = 3.2\times10^{-10}$ m)			$(a = 3.2\times10^{-10}$ m)
NH_4HSO_3	0.000	1.586	1.586	0.000	1.000
	0.001	1.561	1.588	0.002	0.983
	0.002	1.552	1.590	0.004	0.976
	0.003	1.546	1.590	0.004	0.972
	0.004	1.541	1.592	0.006	0.968
	0.005	1.537	1.593	0.007	0.965
		$(a = 3.2\times10^{-10}$ m)			$(a = 3.2\times10^{-10}$ m)
NH_4IO_3	0.000	1.403	1.403	0.000	1.000
	0.001	1.380	1.404	0.001	0.983
	0.002	1.371	1.405	0.002	0.976
	0.003	1.366	1.405	0.002	0.972
	0.004	1.360	1.405	0.002	0.968
	0.005	1.356	1.405	0.002	0.965

TABLE 4.2   *(Continued)*

Electrolyte	c [a]	D_{OF} [b]	F_M [c]	$(\Delta_1 + \Delta_2)$ [c]	F_T [d]
		$(a = 3.2\times10^{-10}$ m$)$			$(a = 3.2\times10^{-10}$ m$)$
NH_4NO_2	0.000	1.936	1.936	0.000	1.000
	0.001	1.907	1.940	0.004	0.983
	0.002	1.896	1.943	0.007	0.976
	0.003	1.889	1.943	0.007	0.972
	0.004	1.882	1.946	0.010	0.967
	0.005	1.877	1.947	0.011	0.964
		$(a = 3.2\times10^{-10}$ m$)$			$(a = 3.2\times10^{-10}$ m$)$
NH_4NO_3	0.000	1.931	1.931	0.000	1.000
	0.001	1.901	1.934	0.003	0.983
	0.002	1.890	1.936	0.005	0.976
	0.003	1.883	1.937	0.006	0.972
	0.004	1.875	1.939	0.008	0.967
	0.005	1.871	1.941	0.010	0.964
		$(a = 3.1\times10^{-10}$ m$)$			$(a = 3.1\times10^{-10}$ m$)$
NH_4OH	0.000	2.858	2.858	0.000	1.000
	0.001	2.808	2.856	−0.002	0.983
	0.002	2.789	2.858	0.000	0.976
	0.003	2.776	2.856	−0.002	0.972
	0.004	2.765	2.856	−0.002	0.968
	0.005	2.755	2.858	0.000	0.964
		$(a = 3.4\times10^{-10}$ m$)$			$(a = 3.4\times10^{-10}$ m$)$
NH_4SCN	0.000	1.861	1.861	0.000	1.000
	0.001	1.833	1.865	0.004	0.983
	0.002	1.823	1.868	0.007	0.976
	0.003	1.815	1.867	0.006	0.972
	0.004	1.809	1.873	0.012	0.966
	0.005	1.803	1.878	0.017	0.960

TABLE 4.2   *(Continued)*

Electrolyte	c [a]	D_{OF} [b]	F_M [c]	$(\Delta_1 + \Delta_2)$ [c]	F_T [d]
		$(a = 4.0\times10^{-10}$ m)			$(a = 4.0\times10^{-10}$ m)
$(NH_4)_2SO_3$	0.000	1.454	1.454	0.000	1.000
	0.001	1.383	1.465	0.011	0.944
	0.002	1.360	1.470	0.016	0.925
	0.003	1.346	1.476	0.022	0.912
	0.004	1.334	1.479	0.025	0.902
	0.005	1.324	1.481	0.027	0.894
		$(a = 3.1\times10^{-10}$ m)			$(a = 3.1\times10^{-10}$ m)
$(NH_4)_2SO_4$	0.000	1.530	1.530	0.000	1.000
	0.001	1.455	1.544	0.014	0.942
	0.002	1.429	1.550	0.020	0.922
	0.003	1.413	1.556	0.026	0.908
	0.004	1.400	1.561	0.031	0.897
	0.005	1.390	1.567	0.037	0.887

[a]c in (mol dm^{-3}) units. [b]D_{OF}, F_M, Δ_1 and Δ_2 in $(10^{-9}$ m^2 s^{-1}) units. [c]$F_M = (D° + \Delta_c + \Delta_a)$, where $(\Delta_1 + \Delta_2)$ represents the electrophoretic correction (Eq. (3)). [d]F_T from available activity coefficient data [6].

In general, for these ammonium salts, when their concentration increase, their diffusion coefficients, D_{OF}, as well as the respective gradient of the free energy, F_T, decrease, and, their mobility factors, F_M, increase. Thus, from these results, we can conclude that the variation in D_{OF} is due mainly to the variation of F_T (attributed to the nonideal thermodynamic behavior), and, to a lesser extent, to the electrophoretic effect in the mobility factor, F_M (Table 4.2), which leads, therefore, to obtaining small values of the sum $(\Delta_c + \Delta_a)$. Therefore, the behavior of diffusion and the mobility and thermodynamic factors of the above systems appear to be affected by the presence of ion-ion interactions (considered in this model), affecting it in two ways: firstly, reducing the activity of the solute as compared with a fully dissociated electrolyte, and consequently leading to lower values of F_T with the concentration and secondly, increasing F_M, due to the aggregate species resulting from these interactions [5] may offer less resistance to motion through the liquid. In fact, if we consider the loss of hydration water from ammonium cation and different anions.

4.5   CONCLUSIONS

In spite of their limitations, the Onsager-Fuoss theory appears to be useful to theo-retically estimate the diffusion coefficients, D_{OF}, of these electrolytes in aqueous solutions and thus, to contribute to a better understanding of the structure of those systems. Because slight variations in this a parameter have little effect on the final results of D_{OF}, the ionic-size values used in Eqs. (1) to (4) were obtained as average ones from those in the literature and ours calculated from MM methods. In addition, for all systems, the variation in the diffusion coefficient, D, is mainly due to the variation of the F_T contribution (attributed to the nonideal thermodynamic behavior) and, to a lesser extent, to the electro phoretic effect in the mobility factor, F_M.

ACKNOWLEDGMENTS

Financial support from FCT (FEDER)-PTDC/AAC-CLI/098308/2008 and PTDC/AAC-CLI/118092/2010 is gratefully acknowledged. One of the authors (A.C.F.R.) is grateful for the *Sabbatical Leave Grant* (BSAB) from *Fundação para a Ciência e Tecnologia*.

KEYWORDS

- **Ammonium salts**
- **Diffusion**
- **Electrolyte solutions**
- **Ion size**

REFERENCES

1. Brugnoni, L. I., Lozano, E. J., & Cubitto, M. A. (2012). *J. Food Process Eng. 35,* 104–119.
2. Beltrame, C. A., Kubiak, G. B., Lerin, L. A., Rottava, I., Mossi, A. J., de Oliveira, D., Cansian, R. L., Treichel, H., & Toniazzo, G. (2012). *Ciencia Tecnol. Aliment,* 228–233.
3. Paolino, D., Muzzalupo, R., Ricciardi, A., Celia, C., Picci, N., & Fresta, M. **(2012).** *Biomed Microdevices, 9,* 421–433.
4. Leclercq, L., Lubart, Q., Dewilde, A., Aubry, J. M., Jean-Marie, Nardello-Rataj, V., & Veronique (2012). *Eur. J. Pharm. Sci., 46,* 336–345.
5. Robinson, R. A., & Stokes, R. H. (1959). Electrolyte Solutions second ed. Butterworths, London.
6. Harned, H. S. B., & Owen, B. (1964). The Physical Chemistry of Electrolytic Solutions, third ed., Reinhold Pub. Corp., New York.

7. Tyrrell, H. J. V., Harris, K. R. (1984). Diffusion in Liquids: a Theoretical and Experimental Study, Butterworths, London.
8. Lobo, V. M. M. (1990). Handbook of Electrolyte Solutions, Elsevier, Amsterdam.
9. Onsager, L., & Fuoss, R. M. (1932) *J. Phys. Chem., 36,* 2689–2778.
10. Lobo, V. M. M., Ribeiro, A. C. F., & Andrade, S. G. C. S. (1996). Port. Electrochim. Acta *14,* 45–124.
11. Lobo, V. M. M., Ribeiro, A. C. F., Andrade, S. G. C. S. (1995). *Ber. Buns. Phys. Chem., 99,* 713–720.
12. Ribeiro, A. C. F., Esteso, M. A., Lobo, V. M. M., Burrows, H. D., Amado, A. M., Amorim da Costa, A. M., & Sobral, A. J. F. N., Azevedo, E. F. G., Ribeiro, M. A. F. (2006). *J. Mol. Liq., 128,* 134–139.
13. Ribeiro, A. C. F., Lobo, V. M. M., Burrows, H. D., Valente, A. J. M., Amado, A.M., Sobral, A. J. F. N., Teles, A. S. N., Santos, C. I. A. V., Esteso, M. A. (2008). *J. Mol. Liq, 140,* 73–77.
14. Ribeiro, A. C. F., Lobo, V. M. M., Burrows, H. D., Valente, A. J. M., Sobral, A. J. F. N., Amado, A. M., Santos, C. I. A. V., & Esteso, M. A. (2009). *J. Mol. Liq., 146,* 69–73.
15. Ribeiro, A. C. F., Barros, M. C. F., Sobral, A. J. F. N., Lobo, V. M. M., & Esteso, M. A. (2010). *J. Mol. Liq., 156,* 124–127.
16. Kielland, J. (1937). *J. Am. Chem. Soc., 59,* 1675–1678.
17. Burkert, U., & Alinger, N. L. (1982). Molecular Mechanics. ACS Monograph 177, American Chemical Society, Washington, DC.
18. Levine, I. N. (1991). Quantum Chemistry, Englewood Cliffs, New Jersey, Prentice Hall, 455–544.

CHAPTER 5

A LECTURE NOTE ON INFLUENCE OF ELECTRIC CURRENT ON THE OXIDATION PROCESS OF CONDUCTIVE POLYMER COMPOSITIONS

E. O. SAZONOVA and I. A. KIRSH

CONTENTS

ABSTRACT

This chapter shows influence of electric current on the oxidation of conductive polymer compositions. A program that describes mathematically the oxidation process of the studied materials in different aging conditions, has been created.

5.1   INTRODUCTION

Recently conductive polymer materials (CPM) have obtained a wide usage in different branches of industry. It takes place due to combination of performance characteristics, availability and low cost. Conductive polymer materials have conductivity specific to metals, and such advantages of the plastics as corrosive resistance, high processing quality, low density, and elasticity. Nowadays the most perspective method of conductive polymer compositions generation is introduction of conductive materials (such as metal powder, graphite, soot) to polymer dielectric [1].

Advantages of charged CPM in comparison with high molecular semiconductors include the possibility of controlling of their conductivity, technological, physical and chemical, and performance properties in accordance with actual practical problems. Variation of performance properties of CPM in a wide range is possible due to rational choosing of the polymer binder and the conductive filler as well as due to optimum ratio polymer filler [2].

At the present time the demand for conductive materials used as heating elements, antistatic coatings, and electrode circuits is increasing. Ever increasing production line gives importance to the problem of functional endurance of such products.

In the process of storage and operation conductive polymer compositions undergo chemical and physical changes resulting in the loss of performance properties. These processes are caused by different factors. One of the key factors reducing efficiency of the products is the process of chemical interaction of polymer materials with atmospheric oxygen. However, in the process of exploitation products of conductive polymer materials undergo exposure to not only atmospheric oxygen, but also such factors as temperature, electric current, humidity, mechanic loading. All of it can result in acceleration of oxidation process of conductive polymer materials and loss of performance characteristics. It has been observed that oxidation process strongly effects CPM properties; thus, in the process of oxidation changing of electrical parameters and reducing of physical and mechanical characteristics take place due to destructive processes taking place during the oxidation of CPM at the simultaneous exposure to stress and temperature [1, 3, 4].

Thus, the aim of this chapter was the study of electric current effect on the oxidation process and properties of CPM as well as development of the model of materials oxidation process under different conditions of aging.

5.2 EXPERIMENTAL PART

Compositions based on synthetic poly isoprene rubber (SKI-3) and poly pyro melt-itimide (PPMI) have been studied. Conducting carbon black used as a filler were Technical carbon (TC) P-234, P-514, P-803 [1, 5].

For the experiment technique of quick aging of CPM was chosen [1]. Oxidation temperature range of CPM based on SKI-3 used was 60°C to140°C, and of CPM based on PPMI 120°C to 240°C. Oxygen pressure in the system was 300 mm Hg. Electrical voltage range 0 to 60 V.

5.3 RESULTS AND DISCUSSION

In Fig. 5.1, pressure variation in the system and aging time of CPM based on SKI-3 containing carbon black P-234 dependence is given.

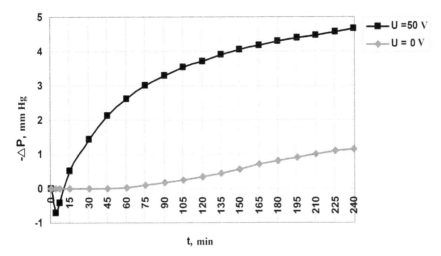

FIGURE 5.1   Pressure variation in system and aging time of CPM based on SKI-3 containing technical carbon P-234 dependence. Ageing conditions: Po2 = 300 mm Hg, T = 100 °C.

At the initial stage of aging of CPM under exposure of current a period of gaseous products evolution is observed. This period is absent in case of oxidation without current exposure.

Similar patterns of CPM oxidation were obtained for other compositions SKI-3 filled with carbon black P-514, P-803 as well as for compositions based on PPMI. It should be noted that with an increase of aging temperature, evolution of gaseous products increases; in case of simultaneous exposure of electrical current and oxygen oxidation rate increases. In case of artificial aging without exposure to electrical current with increasing of the temperature induction period of oxidation reduces and oxidation rate increases; at the same time CPM samples oxidation rate without current exposure is lower than for samples oxidized at electrical current exposure.

Mathematical treatment of kinetic curves of CPM oxidation at different temperatures was carried out. At the curves for CPM samples under exposure to electrical current critical points were chosen: amount of gaseous products, time of evolvement. For the curves of the samples without current exposure induction period was marked.

General view of kinetic curve (Fig. 5.2) is described by the following processes: first type of process (I) is evolvement of gaseous products, second type (II) is absorption of the oxygen by the sample, third type (III) supposes two phenomena: evolvement of gaseous products and oxygen absorption.

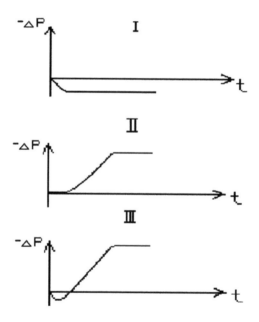

FIGURE 5.2   Stages of CPM oxidation process.

Based on the undertaken study a model of CPM oxidation process was proposed. This model allows divide a view of kinetic curve into two segments: first segment is a period of gaseous products evolvement, reflected on the curve from the beginning of the running time to the point of minimum. This segment is typical only for CPM under exposure to current, because without application of voltage on the sample this segment is absent, and induction period is marked on the curve. Second segment was distinguished from the minimum point in the negative area of kinetic curve to the end point of CPM oxidation time.

Such approach made it possible to determine mathematical formulation of the segment of gaseous products evolvement. With a probability of 0.99 shape of curve stays within quadratic dependence $-\Delta P = at^2 + bt + c$, where a, b, c – coefficients of the equation, ΔP – pressure change in the system, t – time of artificial aging of the CPM (Fig. 5.3).

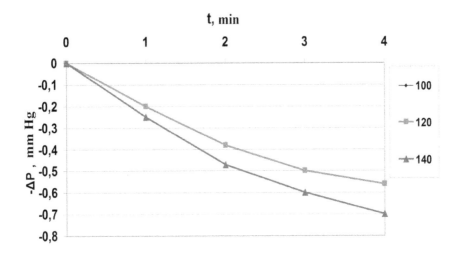

FIGURE 5.3 Dependence of pressure change on aging time for segment of gaseous products evolvement for CPM based on SKI-3 filled with P-234. Conditions of aging: $Po2 = 300$ mm Hg, $U = 60$ V.

Based on experimental data coefficients a, b, c were calculated.

Temperature has a specific impact on the initial period of gaseous products evolvement during CPM oxidation under exposure of electrical current. In Table 5.1 some values of pressure change in the lowest point of the kinetic curve $-\Delta P_{min}$ and time of products evolvement t_{min} is given.

TABLE 5.1   Amount of Evolving Gaseous Products Resulting from CPM Ageing

Composition	Temperature, °C	Pressure change in the lowest point–ΔP_{min}, mm Hg	Time of gaseous products evolvement t_{min}, min
PPMIP-234	200	0.24	30
	300	0.7	27
PPMI P-514	200	0.49	32
	300	0.8	28
PPMI P-803	200	0.46	45
	300	1.35	30
SKI 3 P-214	120	0.7	4
	140	0.56	4.2
SKI 3 P-514	120	0.42	12
	140	0.56	6
SKI 3 P-803	120	0.35	7
	140	0.56	4

As it is seen from given data, with increasing of the oxidation temperature amount of gaseous products increases and period of their evolvement reduces. On the basis of these facts dependences between pressure change in the lowest point of kinetic curve and temperature $\Delta P_{min} = f(T)$ and between time of evolvement and time were plotted.

In Fig. 5.4, dependences between the calculated coefficients and CPM oxidation temperature is given.

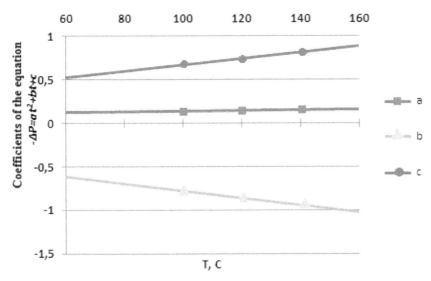

FIGURE 5.4 Dependence of coefficients a, b and c of the gaseous products evolvement equation $\Delta P = at^2 + bt + c$ on temperature for CPM based on SKI-3 filled with P-234. Conditions of aging: $Po_2 = 300$ mm Hg, $U = 60$ V.

Based on these data dependences between coefficients of the equation a, b and c and temperature were obtained:

$a = 0.0115T + 0.1247$
$b = -0.081T - 0.732$
$c = 0.0695T + 0.6073$

Then mathematical analysis of the second segment of the kinctic curve (segment of oxygen absorption by CPM) was performed. In this case it was determined that this process is most likely described by logarithmic equation $-\Delta P = dLn(t) + f$. In Fig. 5.5, dependences between calculated coefficients of this equation and CPM oxidation temperature are shown.

Based on these data dependences between coefficients of the equation d and f and temperature were obtained:

$d = 0.158T + 1.3487$
$f = 0.0293T - 0.528$

Undertaken studies of the artificial aging of CPM have allowed determining that with increasing of the voltage, amount of gaseous products and CPM oxidation rate increase (Fig. 5.6).

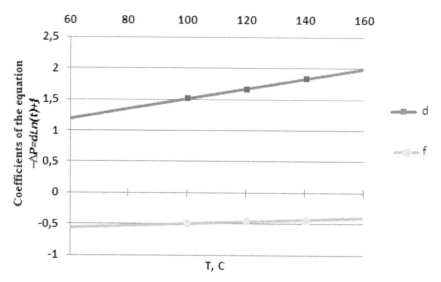

FIGURE 5.5   Dependence of coefficients d and f of the equation of the oxygen absorption segment $-\Delta P = dLn(t) + f$ on temperature for CPM based on SKI-3 filled with P-234. Conditions of aging: $Po_2 = 300$ mm Hg, $U = 60$ V.

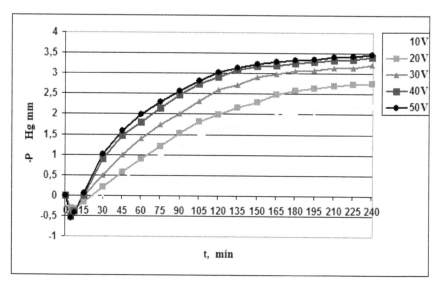

FIGURE 5.6   Dependence of pressure change on aging time for CPM based on SKI-3 depending on voltage. Conditions of aging: $Po_2 = 300$ mm Hg, $T = 100$ °C.

Because kinetic oxidation curves of CPM oxidation at different conditions of voltage exposure (range 10 to 50 V) on the sample looks alike, we performed similar operations to obtain experimental data for CPM aged in the conditions of simultaneous exposure to current and temperature. Equations for the segment of gaseous products evolvement and oxygen absorption segments, coefficients of the equations and their dependences on the voltage for these segments have been determined.

On the basis of these data the program which is aimed at selection of the major oxidation factor for material and allows plotting kinetic oxidation curve for CPM in different conditions of aging has been developed.

The program is carried out after introduction at it is start of two aging factors temperature and voltage. In the table values of pressure change in the system and CPM oxidation time necessary for plotting are output.

This program allows to obtain graphic plot of kinetic oxidation curve $-\Delta P = f(t)$ under exposure to the voltage and temperature in the range 0 to 60 V and 60 to 160 °C after entering two oxidation factors (Temperature and voltage) for CPM based on SKI–3 and PPMI.

This program can be used for calculation of the amount of evolving gaseous products and time of their evolvement, as well as parameters of oxidation process such as maximum oxidation rate.

5.4   CONCLUSIONS

On the basis of the performed experiments influence of the voltage and temperature on the CPM based on PPMI and SKI-3 oxidation process has been determined. Program for data processing allowing simulation of CPM oxidation process with demonstration of the process has been developed.

KEYWORDS

- **Artificial aging**
- **Conductive polymer composition**
- **Oxidation**

REFERENCES

1. Kirsh, I. A. Ph. D thesis in Engineering Science. Effect of Electrical and Mechanical Powers on the kinetics of Poly Isoprene Conductive Composition Oxidation.
2. Gul, V. E., & Shenfil, L. Z. (1984). Conductive polymer compositions. *M.: Chemistry*, 240p.

3. Kirsh, I. A., Gul, V. E., Kvasnikova, E. V., Zabulonov, D. U., & Soina, M. V. (2007). Influence of electric and mechanical forces on the oxidation kinetics and electro conductive poly isopreprene composition properties, *Oxidation communications, 30(1),* 180–184.
4. Kirsh, I. A., Gul,' V. E., Zabulonov, D. U., & Kvasnikova, E. V. (2006). The influence of climatic factors on the properties of the conductive polymer composition. *Plasticheskie massy, 6,* 14–16.
5. Ryzhenkoval, P. Ph. D. thesis in Engineering Science, Influence of electric current on the oxidation of the electroconductive resin composition based on polypropylene and SKI–3.

CHAPTER 6

A TECHNICAL NOTE ON QUANTUM AND WAVE CHARACTERISTICS OF SPATIAL-ENERGY INTERACTIONS

G. A. KORABLEV and G. E. ZAIKOV

CONTENTS

ABSTRACT

It is demonstrated that for two-particle interactions the principle of adding recipro-cals of energy characteristics of subsystems is performed for processes flowing by the potential gradient, and the principle of their algebraic addition for the processes against the potential gradient.

The equation of the dependence of spatial-energy parameter of free atoms on their wave, spectral and frequency characteristics has been obtained.

6.1 INTRODUCTION

Quantum conceptualizations on the composition of atoms and molecules make the foundation of modern natural science theories. Thus, the electronic angular mo-mentum in stationary condition equals the integral multiple from Planck's constant. This main quantum number and three other combined explicitly characterize the state of any atom. The repetition factors of atomic quantum characteristics are also expressed in spectral data for simple and complex structures.

It is known that any periodic processes of complex shape can be shown as sepa-rate simple harmonic waves. "By Fourier theory, oscillations of any shape with pe-riod T can be shown as the total of harmonic oscillations with period's T_1, T_2, T_3, T_4, etc. Knowing the periodic function shape, we can calculate the amplitude and phases of sinusoids, with this function as their total [1].

Therefore, many regularity in intermolecular interactions, complex formation and nano thermodynamics are explained with the application of functional divisible quantum or wave energy characteristics of structural interactions.

In this research we tried to apply the conceptualizations on spatial-energy pa-rameter (P-parameter) for this.

6.1.1 ON TWO PRINCIPLES OF ADDING ENERGY CHARACTERISTICS OF INTERACTIONS

The analysis of the kinetics of various physic-chemical processes demonstrates that in many cases the reciprocals of velocities, kinetic or energy characteristics of the corresponding interactions are added.

Here are some examples: ambipolar diffusion, total rate of topochemical reac-tion, and change in the light velocity when transiting from vacuum into the given medium, effective permeability of bio membranes.

In particular, such assumption is confirmed by the formula of electron transport probability (W_∞) due to the overlapping of wave functions 1 and 2 (in stationary state) during electron-conformation interactions:

$$W_\infty = \frac{1}{2} \frac{W_1 W_2}{W_1 + W_2} \qquad (1)$$

Equation (1) is applied when evaluating the characteristics of diffusion processes accompanied with nonradiating electron transport in proteins [2].

Also: "It is known from the traditional mechanics that the relative motion of two particles with the interaction energy U(r) is the same as the motion of a material point with the reduced mass μ :

$$\frac{1}{\mu} = \frac{1}{m_1} + \frac{1}{m_2} \qquad (2)$$

in the field of central force U(r), and total translational motion – as the free motion of the material point with the mass:

$$m = m_1 + m_2 \qquad (3)$$

Such situation can be also found in quantum mechanics" [3]

The problem of two-particle interactions flowing by the bond line was solved in the time of Newton and Lagrange:

$$\mathring{A} = \frac{m_1 v_1^2}{2} + \frac{m_2 v_2^2}{2} + U(\bar{r}_2 - \bar{r}_1), \qquad (4)$$

where E-system total energy, first and second components – kinetic energies of the particles, third potential energy between particles 1 and 2, vectors $\bar{r}_2$ and $\bar{r}_1$ characterize the distance between the particles in final and initial states.

For moving thermodynamic systems the first law of thermodynamics can be shown as follows [4]:

$$\delta \mathring{A} = d\left(U + \frac{mv^2}{2} \right) \pm \delta A, \qquad (5)$$

where: δE – amount of energy transferred to the system; component $d\left(U + \frac{mv^2}{2} \right)$

characterizes changes in internal and kinetic energies of the system; $+\delta A$ – work performed by the system; $-\delta A$ – work performed on the system.

Since the work numerically equals the change in the potential energy, then:

$$+\delta A = -\Delta U \text{ и } -\delta A = +\Delta U \qquad (6,7)$$

Probably not only the value of potential energy but also its changes are important in thermodynamic and also in many other processes in the dynamics of interactions of moving particles. Therefore, by the analogy with Eq. (4) the following should be fulfilled for two-particle interactions:

$$\delta E = d\left(\frac{m_1 v_1^2}{2} + \frac{m_2 v_2^2}{2}\right) \pm \Delta U \qquad (8)$$

Here

$$\Delta U = U_2 - U_1, \qquad (9)$$

where U_2 and U_1 – potential energies of the system in final and initial states.

At the same time, the total energy (E) and kinetic energy $\left(\dfrac{mv^2}{2}\right)$ can be calculated from their zero value. In this case only the last component is modified in the Eq. (4).

The character of the changes in the potential energy value (ΔU) was analyzed by its index for different potential fields as given in Table 6.1.

From the table it is seen that the values of ΔU and consequently $+\delta A$ (positive work) correspond to the interactions taking place by the potential gradient, and ΔU and $-\delta A$ (negative work) take place during the interactions against the potential gradient.

The solution of two-particle problem of the interaction of two material points with masses m_1 and m_2 obtained under the condition of no external forces available corresponds to the interactions taking place by the gradient, the positive work is performed by the system (similar to attraction process in the gravitation field).

The solution for this equation through the reduced mass (μ) [5] is Lagrangian equation for the relative motion of the isolated system of two interacting material points with masses m_1 and m_2, in coordinate x it looks as follows:

$$\mu \cdot x'' = -\frac{\partial U}{\partial x}; \quad \frac{1}{\mu} = \frac{1}{m_1} + \frac{1}{m_2}.$$

Here: U – mutual potential energy of material points; μ – reduced mass. At the same time $x'' = a$ (characteristic of system acceleration). For elementary regions of interactions Δx can be taken as follows:

$$\frac{\partial U}{\partial x} \approx \frac{\Delta U}{\Delta x} \quad \text{That is: } \mu a \Delta x = -\Delta U. \text{ Then:}$$

$$\text{or: } \frac{1}{\Delta U} \approx \frac{1}{\Delta U_1} + \frac{1}{\Delta U_2} \qquad (10)$$

where ΔU_1 and ΔU_2 – potential energies of material points on the elementary region of interactions, ΔU – resulting (mutual) potential energy of these interactions.

TABLE 6.1   Directedness of Interaction Processes

No	Systems	Potential field type	Process	U	r_2/r_1 (x_2/x_1)	U_2/U_1	Index ΔU	Index δA	Process directedness in the potential field
1	opposite Electric charges	electrostatic	attraction	$-k\dfrac{q_1q_2}{r}$	$r_2<r_1$	$U_2>U_1$	-	+	By gradient
			Repulsion	$-k\dfrac{q_1q_2}{r}$	$r_2>r_1$	$U_2<U_1$	+	-	Against gradient
2	Same electric charges	Electrostatic	Attraction	$k\dfrac{q_1q_2}{r}$	$r_2<r_1$	$U_2>U_1$	+	-	Against gradient
			Repulsion	$k\dfrac{q_1q_2}{r}$	$r_2>r_1$	$U_2<U_1$	-	+	By gradient
3	Elementary masses m_1 and m_2	Gravitational	Attraction	$-\gamma\dfrac{m_1m_2}{r}$	$r_2<r_1$	$U_2>U_1$	-	+	By gradient
			Repulsion	$-\gamma\dfrac{m_1m_2}{r}$	$r_2>r_1$	$U_2<U_1$	+	-	Against gradient
4	Spring deformation	Field of spring forces	Compression	$k\dfrac{\Delta x^2}{2}$	$x_2<x_1$	$U_2>U_1$	+	-	Against gradient
			Stretching	$k\dfrac{\Delta x^2}{2}$	$x_2>x_1$	$U_2>U_1$	+	-	Against gradient
5	Photoeffect	Electrostatic	Repulsion	$k\dfrac{q_1q_2}{r}$	$r_2>r_1$	$U_2<U_1$	-	+	By gradient

Thus,

1. In systems in which the interaction takes place by the potential gradient (positive work), the resultant potential energy is found by the principle of adding the reciprocals of the corresponding energies of subsystems [6]. The reduced mass for the relative motion of isolated system of two particles is calculated in the same way.

2. In systems in which the interaction takes place against the potential gradient (negative work), their masses and corresponding energies of subsystems (similar to Hamiltonian) are added algebraically.

6.2   INITIAL CRITERIA

From the Eq. (10) it is seen that the resultant energy characteristic of the system of interaction of two material points is found by the principle of adding the reciprocals of initial energies of interacting subsystems.

"Electron with the mass m moving near the proton with the mass M is equivalent to the particle with the mass $m_r = \dfrac{mM}{m+M}$ " [7].

Therefore modifying the Eq. (10), we can assume that the energy of atom valence orbitals (responsible for interatomic interactions) can be calculated [6] by the principle of adding the reciprocals of some initial energy components based on the equations:

$$\frac{1}{q^2/r_i} + \frac{1}{W_i n_i} = \frac{1}{P_E} \text{ or } \frac{1}{P_0} = \frac{1}{q^2} + \frac{1}{(Wrn)_i}; \ P_E = P_0/r_i \quad (11), (12), (13)$$

here: W_i – orbital energy of electrons [8]; r_i – orbital radius of i orbital [9]; $q=Z^*/n^*$ – by [10,11], n_i – number of electrons of the given orbital, Z^* and n^* – nucleus effective charge and effective main quantum number, r – bond dimensional characteristics.

P_O is called a spatial-energy parameter (SEP), and P_E – effective P-parameter (effective SEP). Effective SEP has a physical sense of some averaged energy of valence orbitals in the atom and is measured in energy units, for example, in electron-volts (eV).

The values of P_0-parameter are tabulated constants for electrons of the given atom orbital.

For SEP dimensionality:

$$[P_0] = [q^2] = [E] \cdot [r] = [h] \cdot [v] = \frac{kgm^3}{s^2} = Jm,$$

where [E], [h] and [v] – dimensionalities of energy, Plank's constant and velocity.

The introduction of P-parameter should be considered as further development of quasi-classical concepts with quantum-mechanical data on atom structure to obtain the criteria of phase-formation energy conditions. For the systems of similarly charged (e.g., orbitals in the given atom) homogeneous systems the principle of algebraic addition of such parameters is preserved:

$$\sum P_E = \sum (P_0/r_i); \ \sum P_E = \frac{\sum P_0}{r} \qquad (14), (15)$$

or:
$$\Sigma P_0 = P_0' + P_0'' + P_0''' + \dots; \; r\Sigma P_E = \Sigma P_0 \qquad (16), (17)$$

Here P-parameters are summed up by all atom valence orbitals.

To calculate the values of P_E-parameter at the given distance from the nucleus either the atomic radius (R) or ionic radius (r_1) can be used instead of r depending on the bond type.

Let us briefly explain the reliability of such an approach. As the calculations demonstrated the values of P_E-parameters equal numerically (in the range of 2%) the total energy of valence electrons (U) by the atom statistic model. Using the known correlation between the electron density (β) and intraatomic potential by the atom statistic model [12], we can obtain the direct dependence of P_E-parameter on the electron density at the distance r_i from the nucleus.

The rationality of such technique was proved by the calculation of electron density using wave functions by Clementi [13] and comparing it with the value of electron density calculated through the value of P_E-parameter.

6.3   WAVE EQUATION OF P-PARAMETER

To characterize atom spatial-energy properties two types of P-parameters are introduced. The bond between them is a simple one:

$$P_E = \frac{P_0}{R}$$

where R – atom dimensional characteristic. Taking into account additional quantum characteristics of sublevels in the atom, this equation can be written down in coordinate x as follows:

$$\Delta P_E \approx \frac{\Delta P_0}{\Delta x} \;\text{ or }\; \partial P_E = \frac{\partial P_0}{\partial x}$$

where the value ΔP equals the difference between P_0-parameter of i orbital and P_{CD}–countdown parameter (parameter of main state at the given set of quantum numbers).

According to the established [6] rule of adding P-parameters of similarly charged or homogeneous systems for two orbitals in the given atom with different quantum characteristics and according to the energy conservation rule we have:

$$\Delta P_E'' - \Delta P_E' = P_{E,\lambda}$$

where $P_{E,\lambda}$ – spatial-energy parameter of quantum transition.

Taking for the dimensional characteristic of the interaction $\Delta\lambda = \Delta x$, we have:

$$\frac{\Delta P_0''}{\Delta \lambda} - \frac{\Delta P_0'}{\Delta \lambda} = \frac{P_0}{\Delta \lambda} \quad \text{or:} \quad \frac{\Delta P_0'}{\Delta \lambda} - \frac{\Delta P_0''}{\Delta \lambda} = -\frac{P_0 \lambda}{\Delta \lambda}$$

Let us again divide by $\Delta \lambda$ term by term:

$$\left(\frac{\Delta P_0'}{\Delta \lambda} - \frac{\Delta P_0''}{\Delta \lambda} \right) \Big/ \Delta \lambda = -\frac{P_0}{\Delta \lambda^2},$$

where:

$$\left(\frac{\Delta P_0'}{\Delta \lambda} - \frac{\Delta P_0''}{\Delta \lambda} \right) \Big/ \Delta \lambda \sim \frac{d^2 P_0}{d\lambda^2}, \text{ i.e.: } \frac{d^2 P_0}{d\lambda^2} + \frac{P_0}{\Delta \lambda^2} \approx 0$$

Taking into account only those interactions when $2\pi\Delta x = \Delta\lambda$ (closed oscillator), we have the following equation:
$$\frac{d^2 P_0}{dx^2} + 4\pi^2 \frac{P_0}{\Delta \lambda^2} \approx 0$$

Since $\Delta\lambda = \dfrac{h}{mv}$, then: $\dfrac{d^2 P_0}{dx^2} + 4\pi^2 \dfrac{P_0}{h^2} m^2 v^2 \approx 0$

or
$$\frac{d^2 P_0}{dx^2} + \frac{8\pi^2 m}{h^2} P_0 E_k = 0 \qquad (18)$$

where $E_k = \dfrac{mV^2}{2}$ – electron kinetic energy.

Schrodinger equation for the stationery state in coordinate x:

$$\frac{d^2 \psi}{dx^2} + \frac{8\pi^2 m}{h^2} \psi E_k = 0$$

When comparing these two equations we see that P_0-parameter numerically correlates with the value of Ψ-function: $P_0 \approx \Psi$; and is generally proportional to it: $P_0 \sim \Psi$. Taking into account the broad practical opportunities of applying the P-parameter methodology, we can consider this criterion as the materialized analog of Ψ-function [14,15].

Since P_0-parameters like Ψ-function have wave properties, the superposition principles should be fulfilled for them, defining the linear character of the equations of adding and changing P-parameter.

6.4 QUANTUM PROPERTIES OF P-PARAMETER

According to Planck, the oscillator energy (E) can have only discrete values equaled to the whole number of energy elementary portions-quants:

$$nE = h\nu = hc/\lambda \tag{19}$$

where h – Planck's constant, ν – electromagnetic wave frequency, c – its velocity, λ – wavelength, n = 0, 1, 2, 3...

Planck's equation also produces a strictly definite bond between the two ways of describing the nature phenomena – corpuscular and wave.

P_0-parameter as an initial energy characteristic of structural interactions, similarly to the Eq. (19), can have a simple dependence from the frequency of quantum transitions:

$$P_0 \sim \hbar(\lambda\nu_0) \tag{20}$$

where: λ – quantum transition wavelength [16]; $\hbar = h/(2\pi)$; ν_0 – Kaiser, the unit of wave number equaled to 2.9979×10^{10} Hz.

In accordance with Rydberg equation, the product of the right part of this equation by the value $(1/n^2 - 1/m^2)$, where n and m main quantum numbers should result in the constant.

Therefore the following equation should be fulfilled:

$$P_0(1/n_1^2 - 1/m_1^2) = N\hbar(\lambda\nu_0)(1/n^2 - 1/m^2) \tag{21}$$

where the constant N has a physical sense of wave number and for hydrogen atom equals $2 \times 10^2 Å^{-1}$.

The corresponding calculations are demonstrated in Table 6.2. There: $r_i' = 0.5292$ Å – orbital radius of 1S-orbital and $r_i'' = 2^2 \times 0.5292 = 2.118$ Å – the value approximately equaled to the orbital radius of 2S-orbital.

The value of P_0-parameter is obtained from the equation (12), for example, for 1S-2P transition:

$1/P_0 = 1/(13.595 \times 0.5292) + 1/14.394 \rightarrow P_0 = 4.7985$ eVÅ

The value q^2 is taken from Refs. [10, 11], for the electron in hydrogen atom it numerically equals the product of rest energy by the classical radius.

The accuracy of the correlations obtained is in the range of percentage error 0.06 (%), that is, the Eq. (21) is in the accuracy range of the initial data.

In the Eq. (21) there is the link between the quantum characteristics of structural interactions of particles and frequencies of the corresponding electromagnetic waves.

But in this case there is the dependence between the spatial parameters distributed along the coordinate. Thus in P_0-parameter the effective energy is multiplied by the dimensional characteristic of interactions, and in the right part of the Eq. (21) the Kaiser value is multiplied by the wavelength of quantum transition.

In Table 6.2 you can see the possibility of applying the Eq. (21) and for electron Compton wavelength ($\lambda_{\kappa} = 2.4261 \times 10^{-12}$ m), which in this case is as follows:

$$P_0 = 10^7 \hbar (\lambda_{\kappa} v_0) \tag{22}$$

(with the relative error of 0.25%).

Integral-valued decimal values are found when analyzing the correlations in the system "proton-electron" given in Table 6.3:

1. Proton in the nucleus, energies of three quarks $5 + 5 + 7 \approx 17$ (MeV) $\rightarrow P_p \approx$ 17 MeV$\times 0.856 \times 10^{-15}$ m $\approx 14.552 \times 10^{-9}$ eVm. Similarly for the electron $P_e = 0.511$ (MeV)$\times 2.8179 \times 10^{-15}$ m (electron classic radius) $\rightarrow P_e = 1.440 \times 10^{-9}$ eVm.

Therefore:

$$P_p \approx 10 \, P_E \tag{23}$$

2. Free proton $P_n = 938.3$ (MeV)$\times 0.856 \times 10^{-15}$ (m) $= 8.0318 \times 10^{-7}$ eVm. For electron in the atom $P_a = 0.511$ (MeV)$\times 0.5292 \times 10^{-5}$(m) $= 2.7057 \times 10^{-5}$ eVm.

Then:

$$3P_a \approx 10^2 P_n \tag{24}$$

The relative error of the calculations by these equations is found in the range of the accuracy of initial data for the proton ($\delta \approx 1\%$).

From Tables 6.2 and 6.3 we can see that the wave number N is quantized by the decimal principle:

$$N = n10^Z,$$

where n and Z – whole numbers.

Other examples of electrodynamics equations should be pointed out in which there are integral-valued decimal functions, for example, in the formula:

$$4\pi\varepsilon_0 c^2 = 10^7,$$

where ε_0 – electric constant.

In Ref. [17] the expression of the dependence of constants of electromagnetic interactions from the values of electron получено P_e-parameter was obtained:

$$k\mu_0 c = k/(\varepsilon_0 c) = P_e^{1/2}c^2 \approx 10/\alpha \tag{25}$$

Here: $k = 2\pi/\sqrt{3}$; μ_0 – magnetic constant; c – electromagnetic constant; α – fine structure constant.

All the conclusions are based on the application of rather accurate formulas in the accuracy range of initial data.

TABLE 6.2   Quantum Properties of Hydrogen Atom Parameters

Orbitals	W_i (eV)	r_i (Å)	q_i^2 (eVÅ)	P_0 (eVÅ)	$P_0(1/n_1^2{-}1/m_1^2)$ (eVÅ)	N (Å^{-1})	λ (Å)	Quantum transition	$Nh\lambda v_0$ (eVÅ)	$Nh\lambda v_0 \times \times(1/n^2{-}1/m^2)$ (eVÅ)
1S	13.595	0.5292	14.394	4.7985	3.5989	2×10^2	1215	1S-2P	4.7951	3.5963
1S						2×10^2	1025	1S-3P	4.0452	3.5954
1S						2×10^2	912	1S-nP	3.5990	3.5990
2S	3.3988	2.118	14.394	4.7985	3.5990	2×10^2	6562	2S-3P		3.5967
2S						2×10^2	4861	2S-4P		3.5971
2S						2×10^2	3646	2S-nP		3.5973
1S	13.595	0.5292	14.394	4.7985		10^7	2.4263×10^{-2}	–	4.7878	

TABLE 6.3   Quantum Ratios of Proton and Electron Parameters

Particle	E (eV)	r (Å)	$P = Er$ (eVÅ)	Ratio
Free proton	938.3×10^6	0.856×10^{-5}	$8.038\times10^3 = P_n$	
Electron in an atom	0.511×10^6	0.5292	$2.7042\times10^5 = P_a$	$3P_a/P_n \approx 10^2$
Proton in atom nuclei	$(5+5+7)\times10^6 = 17\times10^6$	0.856×10^{-5}	$145.52 = P_p$	
Electron	0.511×10^6	2.8179×10^{-5}	$14.399 = P_e$	$P_p/P_e \approx 10$

6.5   CONCLUSIONS

1. Two principles of adding interaction energy characteristics are functionally defined by the direction of interaction by potential gradient (positive work) or against potential gradient (negative work).

2.   Equation of the dependence of spatial-energy parameter on spectral and frequency characteristics in hydrogen atom has been obtained.

KEYWORDS

- **Equation of the dependence**
- **Frequency characteristics**
- **Hydrogen atom parameters**
- **Interaction energy**
- **Potential gradient**
- **Quantum properties**
- **Spatial-energy parameter**

REFERENCES

1. Gribov, L. A., & Prokofyeva, N. I. (1992). *Basics of Physics,* Vysshaya shkola, M 430p.
2. Rubin, A. B. (1987). Biophysics: *THEORETICAL Biophysics*. Vysshaya shkola M, 319p.
3. Blokhintsev, D. I. (1961). *Basics of Quantum Mechanics*, Vysshaya shkola, M. 512p.
4. Yavorsky, B. M., & Detlaf, A. A. (1968). Reference Book in Physics, Nauka M., 939p.
5. Christy, R. W., & Pytte, A. (1969). *The Structure of Matter: An Introduction to Modern physics*. Translated from English, Nauka, M. 596p.
6. Korablev, G. A. (2005). Spatial-Energy Principles of Complex Structures Formation, Netherlands, Brill Academic Publishers and VSP, 426p (Monograph).
7. Eyring, G., Walter, J., & Kimball, G. (1948). *Quantum Chemistry*, M. F. L., 528p.
8. Fischer, C. F. (1972). *Atomic Data, 4*, 301–399.
9. Waber, J. T., & Cromer, D. T. (1965). *J.Chem. Phys, 42(12),* 4116–4123.
10. Clementi, E., & Raimondi, D. L. (1963). Atomic Screening Constants from S.C.F. functions, 1. *J. Chem. Phys., 38(11)*, 2686–2689.
11. Clementi, E., & Raimondi, D. L. (1967). *J. Chem. Phys., 47(4)*, 1300–1307.
12. Gombash, P. (1951). *Statistic Theory of An Atom and Its Applications*, M. I. L., 398p.
13. Clementi, E. (1965). *J. BMS Res. Develop Suppl., 9(2)*, 76.
14. Korablev, G. A., & Zaikov, G. E. (2006). *J. Applied Polymer Science,* USA, *101(3),* 2101–2107.
15. Korablev, G. A., & Zaikov, G. E. (2009). *Progress on Chemistry and Biochemistry,* Nova Science Publishers, Inc., New York, 355–376.
16. Allen, K. W. (1977). *Astrophysical Values*, Mir, M. 446p.
17. Korablev, G. A. (2010). *Exchange Spatial-Energy Interactions.* Izhevsk Publishing house "Udmurt University," 530p (Monograph).

CHAPTER 7

A TECHNICAL NOTE ON NANOCOMPOSITES POLY (VINYL CHLORIDE)/ORGANOCLAY FLAME-RESISTANCE

I. V. DOLBIN, G. V. KOZLOV, G. E. ZAIKOV, and A. K. MIKITAEV

CONTENTS

ABSTRACT

It has been shown within the framework of strange (anomalous) diffusion conception that instantaneous jumps ("Levy's flights") of combustion front from one region of polymeric material into another increase sharply this material flammability. Distance between nano filler particles decreasing reduces such jumps intensity, increasing thereby material flame-resistance. The fractal time of combustion enhancement results in "Levy's flights" intensification and vice versa.

7.1   INTRODUCTION

At it is known [1, 2], organoclay introduction in polymer increases essentially its flame-resistance. This effect usually is explained by "barrier effect" appearance, that is, organoclay nanoparticles form a kind of barriers, preventing combustion front propagation. In this process structural features of organoclay influence essentially on flame-resistance of nano composites, filled with it. For example, organoclay content increasing results in flame-resistance enhancement and exfoliated organoclay suppresses the ability to combustion more effectively that intercalated one [1, 2]. The purpose of this chapter is the structural analysis of nanocomposites polymer/organoclay flame-resistance enhancement within the framework of fractal analysis.

7.2   EXPERIMENTAL PART

Poly (Vinyl Chloride) Plasticate (PVC) of mark U40–13A of standard recipe 8/2, prepared according to GOST 5962–72, was used. Montmorillonite (MMT), prepared from natural clay according to the technique [3], with the cation exchange capacity of 95 mg equivalent/100 g of clay was used [4].

Nano composites PVC/MMT with organoclay contents of 1–7 mass% were obtained by blending in twin speed blender R600/HC 2500 production of firm "Diosna," the design of which ensures turbulent blending with nano composition homogenization high extent and blowing off with hot air. After plasticate of PVC intensive intermixing with organoclay in a blender at temperature 383–393K up to obtaining high-disperse free-flowing mixture the composition was cooled up to temperature 313K and then it was processed on twin screw extruder Thermo Haake, model Reomex RTW 25/42, production of German Federal Republic, at temperatures 398–423K and screw speed of 48rpm. Testing samples were obtained by casting under pressure method of granulated nanocomposites on a casting machine Test Sample Molding Apparatus RR/TS MP of firm Ray-Ran (Taiwan) at temperature 443K and pressure 12 MPa during 4 min [4].

Flame-resistance (putting out a fire time) is measured on device UL-94 of firm Noselab (Italy) according to GOST 28157–80 [4].

Uniaxial tension mechanical tests have been performed on the samples in the shape of two-sided spade with sizes according to GOST 11262–80. The tests have been conducted on universal testing apparatus Gotech Testing Machine CT–TCS 2000, production of German Federal Republic, at temperature 293K and strain rate ~2×10^{-3} s^{-1} [4].

7.3 RESULTS AND DISCUSSION

The authors [5] formulated the fractional equation of transport processes, having the following form:

$$\frac{\partial^\alpha \psi}{\partial t^\alpha} = \frac{\partial^{2\beta}}{\partial r^{2\beta}}(B\psi),$$

(1)

where $\psi = \psi(t, r)$ is particles distribution function, $\partial^{2\beta}/\partial r^{2\beta}$ is Laplacian in d-dimensional Euclidean space, representing the ratio on generalized transport coefficient and d. The introduction of fractional derivatives $\partial^\alpha/\partial t^\alpha$ and $\partial^{2\beta}/\partial r^{2\beta}$ allows to take into account memory effects (α) and nonlocality effects (β) in the context of a single mathematical formalism [5].

The introduction of fractional derivative $\partial^\alpha/\partial t^\alpha$ in the kinetic equation (the Eq. (1)) also allows taking into account random walks in fractal time (RWFT) a "temporal component" of strange dynamic processes in turbulent mediums [5]. The absence of some appreciable jumps in particles behavior serves as a distinctive feature of RWFT; in addition mean-square displacement grows with t as t^α. Parameter α has the sense of fractal dimension of "active" time, in which real particles walks look as random process: active time interval is proportional to t^α [5].

In its turn, the exponent 2β in the equation (1) takes into account particles instantaneous jumps ("Levy's flights") from one region into another. Thus, the exponents ratio α/β gives the ratio of RWFT and "Levy's flights" contact frequencies. The value α/β is equal to [5]:

$$\frac{\alpha}{\beta} = \frac{d_s}{d}$$

(2)

where d_s is spectral dimension of polymeric material structure, d is dimension of Euclidean space, in which a fractal is considered (it is obvious, that in our case $d=3$).

The structure fractal (Hausdorff) dimension d_f can be determined as follows [6]:

$$d_f = (d-1)(1+\nu)$$

(3)

where ν is Poisson's ratio, estimated by mechanical tests results with the aid of the equation [6]:

$$\frac{\sigma_Y}{E} = \frac{1-2\nu}{6(1+\nu)} \tag{4}$$

where σ_Y is yield stress, E is elasticity modulus.

For linear polymers a macromolecular coil fractal dimension D_f is calculated according to the equation [6]:

$$D_f = \frac{2d_f}{3} \tag{5}$$

Further the value d_s can be determined with the aid of the relationship [6]:

$$D_f = \frac{d_s(d+2)}{2} \tag{6}$$

The parameter α value is calculated as follows [7]:

$$\alpha = 0.5(2-D_f) \tag{7}$$

In case of polymeric material sample combustion this reaction completeness degree Q can be determined as follows. As it is known [7], diffusion process duration τ is calculated according to the equation:

$$\tau = \frac{l_{com}^2}{6D} \tag{8}$$

where l_{com} is polymeric material burning depth, D is diffusivity of this material.

The value Q is determined as the ratio:

$$Q = \frac{2l_{com}}{l}, \tag{9}$$

where l is polymer sample initial thickness, which is equal to 4 mm in the considered case.

Diffusivity coefficient D_m for PVC matrix plasticate can be determined from the equation (8) in supposition, that combustion front achieves the sample middle ($l_{com}=2$ mm) and its achievement time τ is equal to experimental value of putting out a fire time $\tau_p=4.5$ s. Then $D_m=0.148\times10^{-7}$ cm^2/s, that corresponds to experimental data for PVC [7]. The diffusivity coefficient D_n for nanocomposites calculation can be performed within the framework of multi fractal model of gases diffusion according to the equation [7]:

$$D_n = D_m \left(\frac{\alpha_{ac}^n}{\alpha_{ac}^m}\right)^{d_m} \tag{10}$$

where α_{ac}^n and α_{ac}^m are accessible for gases diffusion fractions of nanocomposite and matrix polymer, respectively, d_m is gas-diffusate molecule diameter, which is equal to 3.0Å for O_2 [7].

The relative fraction of accessible for diffusion polymeric material α_{ac} is determined as follows [7]:

$$\alpha_{ac} = 1 - \phi_{cl} - \left(\phi_n + \phi_{if}\right) \qquad (11)$$

where ϕ_{cl}, ϕ_n and ϕ_{if} are relative fractions of local order domains (nano clusters), nano filler and interfacial regions, respectively, through which gases diffusion is not realized in virtue of their dense packing [7].

It is obvious, for matrix PVC plasticate $(\phi_n + \phi_{if}) = 0$. For nanocomposites PVC/MMT $(\phi_n + \phi_{if})$ value is determined with the aid of the following percolation relationship [6]:

$$\frac{E_n}{E_m} = 1 + 11\left(\phi_n + \phi_{if}\right)^{1.7} \qquad (12)$$

where E_n and E_m are elasticity moduli of nanocomposite and matrix polymer, respectively.

The value ϕ_{cl} is estimated as follows [8]:

$$\phi_{cl} = 0.03\left(T_g - T\right)^{0.55} \qquad (13)$$

where T_g and T are glass transition and tests temperatures, respectively. For PVC plasticate $T_g = 348K$ [4], that gives the value $\phi_{cl} = 0.272$ for $T = 293K$.

Further the parameter β can be determined according to the Eq. (2). In Fig. 7.1 the relationship $Q(\beta^3)$ (such correlation type was chosen for its linearization) is adduced, which shows the strong dependence of material combustibility, expressed by the parameter Q, on the exponent β, characterizing combustion front instantaneous jumps ("Levy's flights") from one region of sample into another. So, β doubling results in fivefold nanocomposite volume, subjecting to combustion. The relationship $Q(\beta)$ is described analytically by the following empirical formula:

$$Q = 1.26\beta^3 \qquad (14)$$

Let us consider the structural basis of β change at organoclay contents variation. The distance between organoclay particles λ as the first approximation can be estimated according to the equation [6]:

$$\lambda = \left[\left(\frac{4\pi}{3\varphi_n} \right)^{1/3} - 2 \right] \left(\frac{D_p}{2} \right)$$ (15)

where φ_n is organoclay volume content, D_p is nanofiller particle size, which for organoclay platelet can be estimated as proportional to cube root from its three main sizes product: length, thickness and width, which are equal for exfoliated organoclay to 100, 1 and 35 nm, respectively [6].

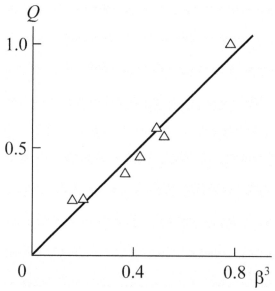

FIGURE 7.1   The dependence of combustion reaction completeness extent Q on exponent b for nanocomposites PVC/MMT.

The value φ_n can be determined according to the well-known equation [6]:

$$\varphi_n = \frac{W_n}{\rho_n}$$ (16)

where W_n is organoclay mass content, ρ_n is its density, which is estimated as follows [6]:

$$\rho_n = 188(D_p)^{1/3}, \text{ kg/m}^3,$$ (17)

where D_p is given in nm.

In Fig. 7.2, the dependence of β on $\lambda^{1/2}$ (such form of the indicated correlation was chosen again with its linearization purpose), which demonstrates β increasing at distance between organoclay particles enhancement and is described analytically by the following empirical equation:

$$\beta = 0.10\lambda^{1/2},\qquad(18)$$

where λ is given in nm.

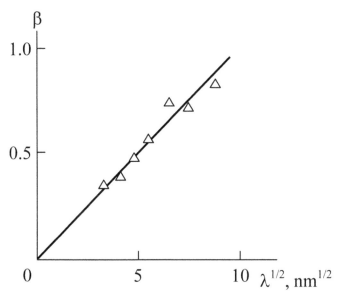

FIGURE 7.2 The dependence of exponent b on distance between nanofiller conditional particles l for nanocomposites PVC/MMT.

The Eqs. (14) and (18) describe theoretically the experimental dependences of nano composites polymer/organoclay combustibility on nano filler structure. Therefore, organoclay contents φ_n increasing results in λ reduction and, respectively, in Q decrease. The transition from organoclay-exfoliated structure up to intercalated one results in a packet (tactoid) from N organoclay platelets formation that increases D_p value and, hence, enhances Q. Both indicated factors influence on β value: λ decreasing results in decay of combustion front jump ("Levy's flight") from one sample region into another, slowing thereby combustion process.

As it has been noted above the interval of active time, that is, material combustion time, is proportional to t^α. This means, that reaction completeness degree Q can be determined as follows [7]:

$$Q \sim t^\alpha\qquad(19)$$

In Fig. 7.3, the dependence of combustion reaction completeness degree Q on its fractal time τ_p^α for nanocomposites PVC/MMT is adduced. As it follows from this figure plot, the indicated dependence is approximated well enough by linear correlation, that is, it corresponds to the relationship (19) and is described by the following empirical equation:

$$Q = 0.47\tau_p^\alpha, \tag{20}$$

where τ_p is given in s.

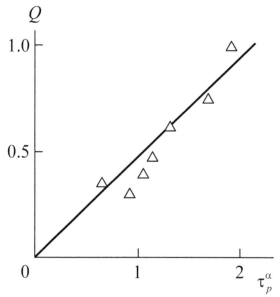

FIGURE 7.3 The dependence of combustion reaction completeness extent Q on fractal time τ_δ^α for nanocomposites PVC/MMT.

The comparison with the similar formula for polymers synthesis [7] shows their principal distinction: the higher (in about 4 times) constant coefficient in the combustion case at the expense of process temperature enhancement. Let us note, that the exponents α and β are interconnected according to the Eq. (2): combustion front jumps ("Levy's flights") intensity reinforcement results in α growth and, hence, to fractal time τ_p^α enhancement and vice versa.

The Eqs. (8) and (10) combination allows to estimates theoretically putting out a fire time τ_p. The theory and experiment comparison in the form of the dependence of τ_p on organoclay mass contents W_n (Fig. 7.4) has shown their good correspondence.

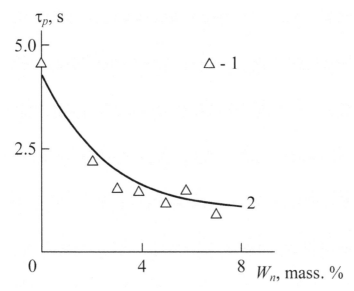

FIGURE 7.4 The experimental (1) and calculated according to the equations (8) and (10) (2) dependences of putting out a fire time tp on organoclay mass contents Wn for nanocomposites PVC/MMT.

7.4 CONCLUSIONS

Thus, this chapter results have shown usefulness of a strange (anomalous) diffusion conception for nanocomposites polymer/organoclay flame-resistance description. Organoclay platelets (tactoids) slow down combustion front jumps ("Levy's flights") from one sample region into another that is physical basis of the barrier effect. The combustion fractal time increasing results in "Levy's flights" intensification and vice versa. The proposed model describes correctly organoclay structure influence on nanocomposites flame-resistance.

KEYWORDS

- **Flame-resistance**
- **Fractal time**
- **Levy's flights**
- **Nanocomposite**
- **Organoclay**
- **Strange diffusion**

REFERENCES

1. Lomakin, S. M., & Zaikov, G. E. (2005). Vysokomolek Soed B (in rus.), 47(1), 104–120.
2. Lomakin, S. M., Dubnikova, I. A., Berezina, S. M., & Zaikov, G. E. (2006). Vysokomolek Soed A (in rus), 48(1), 90–105
3. Clarey, M., Edwards, J., Tzipursky, S. J., Beall, G. W., & Eisenhour, D. D. (2001). Patent 6050509, USA.
4. Khashirova, S. Yu., Borukaev, T. A., Sapaev, Kh., Kh., Ligidov, M. Kh., Kushkhov, Kh. B., & Mikitaev, A. K. (2012). Mater VIII International Sci-Pract Conf. "New Polymer Composite Materials," KBSU, Nal'chik, 218–225.
5. Zelenyi, L. M., & Milovanov, A. V. (2004). Uspekhi Fizicheskikh Nauk, 174(8), 809–852.
6. Mikitaev, A. K., Kozlov, G. V., & Zaikov, G. E. (2008). Polymer Nanocomposites: Variety of Structural Forms and Applications, Nova Science Publishers, Inc., New York, 319.
7. Kozlov, G. V., Zaikov, G. E. & Mikitaev, A. K. (2009). The Fractal Analysis of Gas Transport in Polymers Theory and Practical Applications, Nova Science Publishers Inc. New York, 238.
8. Kozlov, G. V., & Zaikov, G. E. (2004). Structure of the polymer amorphous state. Brill Academic Publishers, Utrecht-Boston, 465.

CHAPTER 8

A SHORT NOTE ON REINFORCEMENT MECHANISM OF POLYMER NANOCOMPOSITES

G. V. KOZLOV, K. S. DIBIROVA, G. M. MAGOMEDOV, and G. E. ZAIKOV

CONTENTS

ABSTRACT

The structural mechanism of polymer nano composites filled with organoclay on supra segmental level was offered. Within the frameworks of these mechanism nanocomposites elasticity modulus is defined by local order domains (nano clusters) sizes similarly to natural nano composites (polymers). Densely packed interfacial regions formation in nano composites at nano filler introduction is the physical basis of nano clusters size decreasing.

8.1   INTRODUCTION

Very often filler (nano filler) is introduced in polymers with the purpose of the latter stiffness increase. This effect is called polymer composites (nano composites) reinforcement and it is characterized by reinforcement degree E_c/E_m (E_n/E_m), where E_c, E_n and E_m are elasticity moduli of composite, nanocomposite and matrix polymer, accordingly. The indicated effect significance results to a large number of quantitative models developments, describing reinforcement degree: micromechanical [1], percolation [2] and fractal [3] ones. The principal distinction of the indicated models is the circumstance that the first ones take into consideration the filler (nano filler) elasticity modulus and the last two don't. The percolation [2] and fractal [3] models of reinforcement assume that the filler (nano filler) role comes to modification and fixation of matrix polymer structure. Such approach is obvious enough, if to account for the difference of elasticity modulus of filler (nano filler) and matrix polymer. So, for the considered in this chapter nanocomposites low density polyethylene/ Na^+-montmorillonite the matrix polymer elasticity modulus makes up 0.2 GPa [4] and nanofiller 400–420 GPa [5], that is, the difference makes up more than three orders. It is obvious, that at such conditions organoclay strain is equal practically to zero and nano composites behavior in mechanical tests is defined by polymer matrix behavior.

Lately it was offered to consider polymers amorphous state structure as a natural nanocomposite [6]. Within the frameworks of cluster model of polymers amorphous state structure it is supposed, that the indicated structure consists of local order domains (clusters), immersed in loosely packed matrix, in which the entire polymer free volume is concentrated [7, 8]. In its turn, clusters consist of several collinear densely packed statistical segments of different macromolecules, that is, they are an amorphous analog of crystallites with stretched chains. It has been shown [9] that clusters are nanoworld objects (true nanoparticles-nano clusters) and in case of polymers representation as natural nanocomposites they play nanofiller role and loosely packed matrix-nanocomposite matrix role. It is significant that the nanoclusters dimensional effect is identical to the indicated effect for particulate filler in polymer nano composites sizes decrease of both nano clusters [10] and disperse particles [11] results to sharp enhancement of nanocomposite reinforcement degree (elasticity modulus). In connection with the indicated observations the question arises: how organoclay introduction in polymer matrix influences on nano clusters size and how the variation of the latter influences on nanocomposite elasticity modulus value. The purpose of this chapter is these two problems solution on the example of nanocomposite linear low-density polyethylene/Na⁺-montmorillonite [4].

8.2 EXPERIMENTAL PART

Linear low density polyethylene (LLDPE) of mark Dowlex-2032, having melt flow index 2.0 g/10 min and density 926 kg/m³ that corresponds to crystallinity degree of 0.49, used as a matrix polymer. Modified Na⁺-montmorillonite (MMT), obtained by cation exchange reaction between MMT and quaternary ammonium ions, was used as nano filler MMT contents makes up 1–7 mass% [4].

Nano composites linear low-density polyethylene/Na⁺-montmorillonite (LLDPE/MMT) were prepared by components blending in melt using Hake twin-screw extruder at temperature 473 K [4].

Tensile specimens were prepared by injection molding on Arburg Allounder 305–210–700 molding machine at temperature 463K and pressure 35 MPa. Tensile tests were performed by using tester Instron of the model 1137 with direct digital data acquisition at temperature 293 K and strain rate ~ 3.35×10^{-3} s⁻¹. The average error of elasticity modulus determination makes up 7%, yield stress 2% [4].

8.3 RESULTS AND DISCUSSION

For the solution of the first from the indicated problems the statistical segments number in one nano cluster n_{cl} and its variation at nanofiller contents change should be estimated. The parameter n_{cl} calculation consistency includes the following stages. At first the nanocomposite structure fractal dimension d_f is calculated according to the equation [12]:

$$d_f = (d-1)(1+v),$$ (1)

where d is dimension of Euclidean space, in which a fractal is considered (it is obvious, that in our case $d=3$), v is Poisson's ratio, which is estimated according to mechanical tests results with the aid of the relationship [13]:

$$\frac{\sigma_Y}{E_n} = \frac{1-2v}{6(1+v)},$$ (2)

where σ_Y and E_n are yield stress and elasticity modulus of nanocomposite, accordingly.

Then nano clusters relative fraction φ_{cl} can be calculated by using the following equation [8]:

$$d_f = 3 - 6\left(\frac{\phi_{cl}}{C_\infty S}\right)^{1/2},$$

(3)

where C_∞ is characteristic ratio, which is a polymer chain statistical flexibility indicator [14], S is macromolecule cross-sectional area.

The value C_∞ is a function of d_f according to the relationship [8]:

$$C_\infty = \frac{2d_f}{d(d-1)(d-d_f)} + \frac{4}{3}.$$

(4)

The value S for low-density polyethylenes is accepted equal to 14.9 Å² [15]. Macromolecular entanglements cluster network density v_{cl} can be estimated as follows [8]:

$$v_{cl} = \frac{\phi_{cl}}{C_\infty l_0 S},$$

(5)

where l_0 is the main chain skeletal bond length which for polyethylenes is equal to 0.154 nm [16].

Then the molecular weight of the chain part between nanoclusters M_{cl} was determined according to the equation [8]:

$$M_{cl} = \frac{\rho_p N_A}{v_{cl}},$$

(6)

where ρ_p is polymer density, which for the studied polyethylenes is equal to ~ 930 kg/m³, N_A is Avogadro number.

And at last, the value n_{cl} is determined as follows [8]:

$$n_{cl} = \frac{2M_e}{M_{cl}},$$

(7)

Where M_e is molecular weight of a chain part between entanglements traditional nodes ("binary hookings"), which is equal to 1390 g/mole for low-density polyethylenes [17].

In Fig. 8.1, the dependence of nanocomposite elasticity modulus E_n on value n_{cl} is adduced, from which E_n enhancement at n_{cl} decreasing follows. Such behavior of nanocomposites LLDPE/MMT is completely identical to the behavior of both particulate-filled [11] and natural [10] nanocomposites.

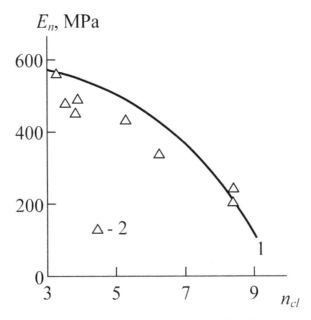

FIGURE 8.1 The dependences of elasticity modulus E_n on Statistical segments number per one nanocluster n_{cl} for nanocomposites LLDPE/MMT. (i) Calculation according to the Eq. (8); (ii) the experimental data.

In Ref. [18] the theoretical dependences of E_n as a function of cluster model parameters for natural nano composites was obtained:

$$E_n = c\left(\frac{\phi_{cl} V_{cl}}{n_{cl}}\right),\qquad (8)$$

where c is constant, accepted equal to 5.9×10^{-26} m^3 for LLDPE.

In Fig. 8.1, the theoretical dependence $E_n(n_{cl})$, calculated according to the Eq. (8), for the studied nanocomposites is adduced, which shows a good enough correspondence with the experiment (the average discrepancy of theory and experiment makes up 11.6%, that is comparable with mechanical tests experimental error). Therefore, at organoclay mass contents W_n increasing within the range of 0–7 mass % n_{cl} value reduces from 8.40 up to 3.17, that is accompanied by nanocomposites LLDPE/MMT elasticity modulus growth from 206 up to 569 MPa.

Let us consider the physical foundations of n_{cl} reduction at W_n growth. The main equation of the reinforcement percolation model is the following one [9]:

$$\frac{E_n}{E_m} = 1 + 11\left(\phi_n + \phi_{if}\right)^{1.7},$$ (9)

where ϕ_n and ϕ_{if} are relative volume fractions of nanofiller and interfacial regions, accordingly.

The value ϕ_n can be determined according to the equation [5]:

$$\phi_n = \frac{W_n}{\rho_n},$$ (10)

where ρ_n is nanofiller density, which is equal to ~ 1700 kg/m³ for Na⁺-montmorillonite [5].

Further the Eq. (9) allows to estimate the value ϕ_{if}. In Fig. 8.2 the dependence $n_{cl}(\phi_{if})$ for nanocomposites LLDPE/MMT is adduced. As one can see, n_{cl} reduction at ϕ_{if} increasing is observed, that is, formed on organoclay surface densely packed (and, possibly, subjecting to epitaxial crystallization [9]) interfacial regions as if pull apart nanoclusters, reducing statistical segments number in them. As it follows from the Eqs. (9) and (8), these processes have the same direction, namely, nanocomposite elasticity modulus increase.

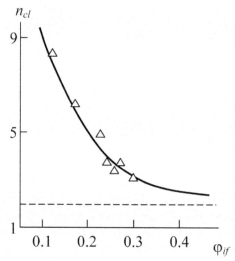

FIGURE 8.2   The dependence of statistical segments number per one nano cluster n_{cl} on interfacial regions relative fraction j_{if} for nano composites LLDPE/MMT. Horizontal shaded line indicates the minimum value $n_{cl} = 2$.

8.4 CONCLUSIONS

Hence, this chapter results demonstrated common reinforcement mechanism of natural and artificial (filled with inorganic nanofiller) polymer nanocomposites. The statistical segments number per one nanocluster reduction at nanofiller contents growth is such a mechanism on suprasegmental level. The indicated effect physical foundation is the densely packed interfacial regions formation in artificial nanocomposites.

KEYWORDS

- **Elasticity modulus**
- **Interfacial regions**
- **Nanoclusters**
- **Nanocomposite**
- **Organoclay**
- **Polyethylene**

REFERENCES

1. Ahmed, S., & Jones, F. R. (1990). *J. Mater. Sci., 25(12),* 4933–4942.
2. Bobryshev, A. N., Kozomazov, V. N., Babin, L. O., & Solomatov, V. I. (1994). Synergetics of Composite Materials. Lipetsk, NPO ORIUS, 154p.
3. Kozlov, G. V., Yanovskii, Yu. G., & Zaikov, G. E. (2010). *Structure and Properties of Particulate-Filled Polymer Composites: the Fractal Analysis.* New York, Nova Science Publishers, Inc., 282p
4. Hotta, S., & Paul, D. R. (2004). *Polymer, 45(21),* 7639–7654.
5. Sheng, N., Boyce, M. C., Parks, D. M., Rutledge, G. C., Abes, J. I., & Cohen, R. E. (2004). *Polymer, 45(2),* 487–506.
6. Bashorov, M. T., Kozlov, G. V., & Mikitaev, A. K. (2009). *Material Ovedenie, 9,* 39–51.
7. Kozlov, G. V., & Novikov, V. U. (2001). Uspekhi Fizicheskikh Nauk, *171(7),* 717–764.
8. Kozlov, G. V., & Zaikov, G. E. (2004). *Structure of the Polymer Amorphous State,* Utretch Boston Brill Academic Publishers, 465p.
9. Mikitaev, A. K., Kozlov, G. V., & Zaikov, G. E. (2008). *Polymer Nanocomposites: Variety of Structural Forms and Applications.* New York, Nova Science Publishers, Inc., 319p.
10. Kozlov, G. V., & Mikitaev, A. K. (2010). *Polymers as natural nanocomposites. Unrealized Potential.* Saarbrücken Lambert Academic Publishing, 323p.
11. Edwards, D. C. (1990). *J. Mater. Sci., 25(12),* 4175–4185.
12. Balankin, A. S. (1991). Synergetics of Deformable Body Moscow, Publishers of Ministry Defence SSSR, 404p.

13. Kozlov, G. V., & Sanditov, D. S. (1994). *An Harmonic Effects and Physical-Mechanical Properties of Polymers*. Novosibirsk, Nauka, 261p.
14. Budtov, V. P. (1992). *Physical Chemistry of Polymer Solutions,* Sankt-Peterburg Khimiya, 384p.
15. Aharoni, S. M. (1985). *Macromolecules, 18(12),* 2624–2630.
16. Aharoni, S. M. (1983). *Macromolecules, 16(9),* 1722–1728.
17. Wu, S. (1989). *J. Polymer Sci.: Part B: Polymer Phys., 27(4),* 723–741.
18. Kozlov, G. V. (2011). *Recent Patents on Chemical Engineering, 4(1),* 53–77.

CHAPTER 9

A RESEARCH NOTE ON REINFORCEMENT DEGREE OF SEMICRYSTALLINE POLYMERS AS NATURAL HYBRID NANOCOMPOSITES

G. M. MAGOMEDOV, K. S. DIBIROVA, G. V. KOZLOV, and G. E. ZAIKOV

CONTENTS

ABSTRACT

It has been shown that at semi crystalline polymers with devitrificated amorphous phase consideration as natural hybrid nanocomposites their anomalous high reinforcement degree is realized at the expense of crystallites partial recrystallization (mechanical disordering), that means crystalline phase participation in these polymers elastic properties formation. It is obvious, that the proposed mechanism is inapplicable for the description of polymer nano composites with inorganic nano fillers.

9.1   INTRODUCTION

As it is known [1], semi crystalline polymers, similar to widely applicable polyethylene and polypropylene, at temperatures of the order of room ones have devitrificated amorphous phase. This means that such phase elasticity modulus is small and makes up the value of the order of 10 MPa [2]. At the same time elasticity modulus of the indicated above polymers can reach values of ~ 1.0–1.4 GPa and is a comparable one with the corresponding parameter for amorphous glassy polymers. In case of the latters it has been shown [3, 4], that they can be considered as natural nano composites, in which local order domains (nano clusters) serve as nano filler and as loosely packed matrix of polymer within the framework of the cluster model of polymers amorphous state structure [5] is considered as matrix. In this case elasticity modulus of glassy loosely packed matrix makes up the value of order of 0.8GPa and a corresponding parameter for polymer (for example, polycarbonate or polyacrylate) ~1.6GPa. In other words, the reinforcement degree of loosely packed matrix by nanoclusters for amorphous glassy polymers is equal to ~2, where as for the indicated above semi crystalline polymers it can exceed two orders. By analogy with amorphous [3, 4] and cross-linked [6] polymers semi crystalline polymers can be considered as natural hybrid nano composites, in which rubber-like matrix is reinforced by two kinds of nano filler: nano clusters (analog of disperse nano filler with particles size of the order of ~ 1 nm [5]) and crystallites (analog of organoclay with platelets size of the order of ~ 30–50 nm [7]). The clarification of abnormally high reinforcement degree mechanism allows to gives an answer to the question, would this mechanism be applicable to polymer nanocomposites, filled with inorganic nano filler (for example, organoclay). Therefore, the purpose of this chapter is the study of reinforcement mechanism of rubber-like matrix of high-density polyethylene (HDPE) at its consideration as natural hybrid nanocomposite.

9.2   EXPERIMENTAL PART

The gas-phase HDPE of industrial production of mark HDPE-276, GOST 16338–16385 with average weight molecular mass $1.4 \ 10^5$ and crystallinity degree 0.723, measured by the sample density, was used.

The testing specimens were prepared by method of casting under pressure on a casting machine Test Samples Molding Apparate RR/TS MP of firm Ray-Ran (Taiwan) at material cylinder temperature 473 K, compression mold temperature 333 K and pressure of blockage 8 MPa.

The impact tests have been performed by using a pendulum impact machine on samples without a notch according to GOST 4746–4780, type II, within the testing temperatures range T=213–333K. Pendulum impact machine was equipped with a piezoelectric load sensor, that allows to determine elasticity modulus E and yield stress σ_Y in impact tests according to the techniques [8] and [9], respectively.

Uniaxial tension mechanical tests have been performed on the samples in the shape of two-sided spade with sizes according to GOST 11262–11280. The tests have been conducted on a universal testing apparatus Gotech Testing Machine CT-TCS 2000, production of German Federal Republic, within the testing temperatures range T=293–363K and strain rate of 2×10^{-3} s^{-1}.

9.3   RESULTS AND DISCUSSION

In Fig. 9.1, the temperature dependences of elasticity modulus E for the studied HDPE have been adduced. As one can see, at comparable testing temperatures E value in case of quasistatic tests is about twice smaller, than in impact ones. Let us note that this distinction is not due to tests type. As it has been noted above, for HDPE with the same crystallinity degree at T=293 K E value can reach 1252 MPa [10]. Let us consider the physical grounds of this discrepancy. The value of fractal dimension d_f of polymer structure, which is its main characteristic, can be determined by several methods application. The first from them uses the following equation [11]:

$$d_f = (d-1)(1+v), \tag{1}$$

where d is dimension of Euclidean space, in which a fractal is considered (it is obvious, that in our case d=3), v is Poisson's ratio, estimated according to the mechanical tests results with the aid of the equation [12]:

$$\frac{\sigma_Y}{E} = \frac{1-2v}{6(1+v)}. \tag{2}$$

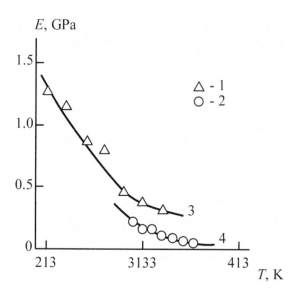

FIGURE 9.1 The dependences of elasticity modulus E on testing temperature T for HDPE, obtained in impact (1, 3) and quasistatic (2, 4) tests. 1, 2 the experimental data, 3, 4 calculations according to the equation (12).

The second method assumes the value d_f calculation according to the equation [5]:

$$d_f = d - 6.44 \times 10^{-10} \left(\frac{\phi_{cl}}{C_\infty S} \right)^{1/2}, \qquad (3)$$

where ϕ_{cl} is relative fraction of local order domains (nanoclusters), C_∞ is characteristic ratio, S is cross-sectional area of macromolecule.

For HDPE $C_\infty = 7$ [13], $S = 14.4$ Å^2 [14] and ϕ_{cl} value can be calculated according to the following percolation relationship [5]:

$$\phi_{cl} = 0.03(1 - K)(T_m - T)^{0.55}, \qquad (4)$$

where K is crystallinity degree, T_m and T are melting and testing temperatures, respectively. For HDPE $T_m \approx 400K$ [15].

And at last, for semi crystalline polymers d_f value can be evaluated as follows [16]:

$$d_f = 2 + K \qquad (5)$$

In Table 9.1, the comparison of d_f values, determined by the three indicated methods has been adduced (K change with temperature was estimated according to the data of Ref. [7]). As one can see, if in case of impact tests the calculation according to all three indicated methods gives coordinated results, then for quasistatic tests estimation according to the Eqs. (1) and (2) gives clearly understated d_f values, especially with appreciation of possible variation of this dimension for nonporous solids ($2 \leq d_f \leq 2.95$ [11]).

TABLE 9.1 The Values of Fractal Dimension d_f of HDPE Structure, Calculated by Different Methods

Tests type	T, K	d_f, the equation (1)	d_f, the equation (3)	d_f, the equation (5)
Quasistatic	293	2.302	2.800	2.723
	303	2.296	2.796	2.723
	313	2.272	2.802	2.713
	323	2.353	2.801	2.693
	333	2.248	2.799	2.673
	343	2.182	2.799	2.663
	353	2.170	2.800	2.643
	363	2.078	2.808	2.633
Impact	213	2.764	2.734	2.723
	233	2.762	2.741	2.723
	253	2.700	2.727	2.723
	273	2.750	2.756	2.723
	293	2.680	2.729	2.723
	313	2.624	2.743	2.713
	333	2.646	2.766	2.763

Let us consider the causes of the indicated discrepancy in more details. At present it has been assumed [1], that in case of semi crystalline polymers with devitrificated amorphous phase deformation in elasticity region, that is, at E value determination, the indicated amorphous phase is only deformed, that defines smaller values of both E and d_f. This conclusion is confirmed by disparity between d_f values, calculated on the basis of mechanical characteristics (the Eqs. (1) and (2)) and crystallinity degree (the Eq. (5)). And on the contrary, a good correspondence of d_f values, obtained by the three indicated methods (Table 9.1), assumes crystalline phase participance at HDPE deformation in elasticity region in case of impact tests (Fig. 9.1).

In Fig. 9.2, the temperature dependence of yield stress σ_Y has been adduced for HDPE in case of both types of tests.

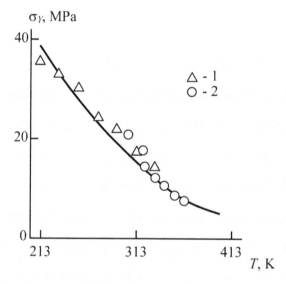

FIGURE 9.2   The dependence of yield stress s_Y on testing temperature T for HDPE, obtained in impact (1) and quasistatic (2) tests.

As one can see, the data for quasistatic and impact tests are described by the same curve. As it has been shown in Ref. [5], σ_Y values are defined by contribution of both nano clusters (i.e., amorphous phase), and crystallites. Combined consideration of the plots of (Figs. 9.1 and 9.2) demonstrates, that the distinction of d_f values, determined according to the Eqs. (1), (2) and (5), is only due to the indicated above structural distinction of HDPE deformation in an elasticity region.

The quantitative evaluation of crystalline regions contribution in HDPE elasticity can be performed within the framework of yield fractal conception [17], according to which the value of Poisson's ration in yield point v_Y can be estimated as follows:

$$v_Y = v\chi + 0.5(1 - \chi), \qquad (6)$$

where v is Poisson's ratio in elastic strains region, determining according to the equation (2), χ is a relative fraction of elastically deformed polymer.

For amorphous glassy polymers it has been shown that χ value is equal to a relative fraction of loosely packed matrix $\varphi_{l.m.}$ In case of semi crystalline polymers in deformation process partial recrystallization (mechanical disordering) of a crystallites part can be realized, the relative fraction of which χ_{cr} is determined by the following equation [5]:

$$\chi_{cr} = \chi - (1 - K) \tag{7}$$

If to consider HDPE as natural hybrid nanocomposite, then amorphous phase (the indicated nanocomposite matrix) elasticity modulus E_{am} can be determined within the framework of high-elasticity conception, using the known equation [2]:

$$G_{am} = kNT, \tag{8}$$

where G_{am} is a shear modulus of amorphous phase, k is Boltzmann constant, N is a number of active chains of polymer network.

As it is known [5], in amorphous phase two types of macromolecular entanglements are present: traditional macromolecular "binary hooking" and entanglements, formed by nano clusters, networks density of which is equal to v_e and v_{cl}, respectively. v_e value is determined within the framework of rubber high-elasticity conception [2]:

$$v_e = \frac{\rho_p N_A}{M_e}, \tag{9}$$

where ρ_p is polymer density, N_A is Avogadro number, M_e is molecular weight of polymer chain part between macromolecular "binary hooking," which is equal to 1390 [18] and ρ_p value can be accepted equal to 960 kg/m³ [15].

In its turn, v_{cl} value can be determined according to the following equation [5]:

$$v_{cl} = \frac{\phi_{cl}}{C_\infty l_0 S}, \tag{10}$$

where l_0 is length of the main chain skeletal bond, for HDPE equal to 1.54Å [13].

E_{am} and G_{am} are connected by a simple fractal formula [11]:

$$E_{am} = d_f G_{am}. \tag{11}$$

Calculation according to the equations (6) and (7) has shown that in case of HDPE quasistatic tests χ_{cr} value is close to zero and in case of impact tests $\chi_{cr}=0.400–0.146$ within the range of $T=231–333K$. Now reinforcement degree of HDPE, considered as hybrid nanocomposite, can be expressed as the ratio E/E_{am}. In Fig. 9.3, the dependence $E/E_{am}(\chi_{cr}^2)$ has been adduced (such form of dependence was chosen for its linearization). As one can see, this linear dependence demonstrates E/E_{am} (or E) increase at χ_{cr} growth and is described analytically by the following relationship:

$$\frac{E}{E_{am}} = 590\chi_{cr}^2, \quad \text{MPa.} \tag{12}$$

The comparison of experimental E and calculated according to the equation (12) E^T elasticity modulus values for the studied HDPE has been adduced in Fig. 9.1. As one can see, the good correspondence between theory and experiment is obtained (the average discrepancy between E and E^T does not exceed 6%, that is, comparable with an error of elasticity modulus experimental determination).

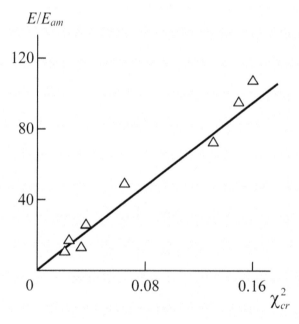

FIGURE 9.3 The dependence of reinforcement degree E/E_{am} on relative fraction of crystalline phase χ_{cr}, subjecting to partial recrystallization, for HDPE in impact tests.

Thus, the performed estimations demonstrated, that high values of reinforcement degree E/E_{am} for semi crystalline polymers, considered as hybrid nano composites (in case of studied HDPE E/E_{am} value is varied within the limits of 10–110) were due to recrystallization process (mechanical disordering of crystallites) in elastic deformation process and, as consequence, to contribution of crystalline regions in polymers elastic properties formation. It is obvious, that this mechanism does not work in case of inorganic nano filler (e.g., organoclay). Besides, a nano filler (crystallites) is formed spontaneously in a polymer crystallization process, that automatically cancels the problem of its dispersion at large K of the order of 70 mass%, whereas to obtain exfoliated organoclay at contents larger than 3 mass% is difficult [19]. Taking into consideration of the above indicated factors it becomes clear, why nanocomposites polymer/organoclay maximum reinforcement degree does not exceed 4 [20].

9.4 CONCLUSIONS

The performed analysis has shown that at the consideration of semi crystalline polymers with devitrificated amorphous phase as natural hybrid nano composites their abnormally high reinforcement degree is realized at the expense of crystallites partial recrystallization (mechanical disordering), that means crystalline phase participation in the formation of these polymers elastic properties. It is obvious, that the proposed mechanism is inapplicable for the description of reinforcement of polymer nano composites with inorganic nano filler.

KEYWORDS

- **Fractal dimension**
- **Natural nanocomposite**
- **Recrystallization**
- **Reinforcement degree**
- **Semicrystalline polymer**

REFERENCES

1. Narisawa, I. (1987). *Strength of Polymeric Materials*, Chemistry Publishing House (in Rus.), Moscow, 400.
2. Bartenev, G. M., & Frenkel, S. Ya. (1990). *Physics of Polymers*, Chemistry Publishing House (in Rus.), Moscow, 432.
3. Kozlov, G. V. (2011). *Recent Patents on Chemical Engineering, 4(1)*, 53–77.
4. Kozlov, G. V., & Mikitaev, A. K. (2010). *Polymers as Natural Nano composites Unrealized Potential*. Lambert Academic Publishing, Saarbrücken, 323.
5. Kozlov, G. V., Ovcharenko, E. N., & Mikitaev, A. K. (2009). *Structure of Polymer Amorphous State* (in rus.). RKhTU Publishing House, Moscow, 392.
6. Magomedov, G. M., Kozlov, G. V., & Zaikov, G. E. (2011). *Structures and Properties of Cross-Linked Polymers,* Smithers Group Company Shaw bury 492.
7. Tanabe, Y., Strobl, G. R., & Fisher, E. W. (1986). *Polymer, 27(8)*, 1147–1153.

8. Kozlov, G. V., Shetov, R. A., & Mikitaev, A. K. (1987). *Russian Polymer Science* (in rus.), *29(5)*, 1109–1110.

9. Kozlov, G. V., Shetov, R. A., & Mikitaev, A. K. (1987). *Russian Polymer Science* (in rus.), *29(9),* 2012–2013.

10. Pegoretti, A., Dorigato, A. & Penati, (2007). A. *EXPRESS Polymer Lett., 1(3),* 123–131.

11. Balankin, A. S. (1991). *Synergetics of deformable Body.* Ministry of Defence SSSR Publishing House, Moscow, 404.

12. Kozlov, G. V., & Sanditov, D. S. (1994). *An harmonic effects and Physical-Mechanical Properties of Polymers,* Nauka (Science) Publishing House (in Rus.), Novosibirsk, 264.

13. Aharoni, S. M. (1983). *Macromolecules, 16(9), 1722–1728.*

14. Aharoni, S. M. (1985). *Macromolecules, 18(12),* 2624–2630.

15. Kalinchev, E. L., & Sakovtseva, M. B. (1983). *Properties and Processing of Thermoplastics* (in Rus.), Chemistry Publishing House, Leningrad, 288.

16. Aloev, V. Z., & Kozlov, G. V. (2002). *Physics of Orientation Phenomena in Polymeric Materials* (in Rus.). Polygraph service and T, Nal'chik, 288.

17. Balankin, A. S., & Bugrimov, A. L. (1992). *Russian Polymer Science* (in rus.), *34(5),* 129–132.

18. Graessley, W. W., & Edwards, S. F. (1981). *Polymer, 22(10),* 1329–1334.

19. Miktaev, A. K., Kozlov, G. V., & Zaikov, G. E. (2008). *Polymer Nano composites: Variety of Structural Forms and Applications.* Nova Science Publishers Inc., New York, 319.

20. Liang, Z. M., Yin, J., Wu, J. H., Qiu, Z. X., & He, F. F. (2004). *Europe Polymer J., 40(2),* 307–314.

CHAPTER 10

KINETIC MODELING OF POLYSTYRENE CONTROLLED RADICAL POLYMERIZATION

NIKOLAI V. ULITIN, ALEKSEY V. OPARKIN, RUSTAM YA. DEBERDEEV, EVGENII B. SHIROKIH, and GENNADY E. ZAIKOV

CONTENTS

ABSTRACT

The kinetic modeling of styrene controlled radical polymerization, initiated by 2,2'-asobis(isobutirnitrile) and proceeding by a reversible chain transfer mechanism was carried out and accompanied by "addition-fragmentation" in the presence dibenzyltritiocarbonate. An inverse problem of determination of the unknown temperature dependences of single elementary reaction rate constants of kinetic scheme was solved. The adequacy of the model was revealed by comparison of theoretical and experimental values of polystyrene molecular-mass properties. The influence of process controlling factors on polystyrene molecular-mass properties was studied using the model.

10.1   INTRODUCTION

The controlled radical polymerization is one of the most developing synthesis methods of narrowly dispersed polymers nowadays [1–3]. Most considerations were given to researches on controlled radical polymerization, proceeding by a reversible chain transfer mechanism and accompanied by "addition-fragmentation" (RAFT reversible addition-fragmentation chain transfer) [3]. It should be noted that for classical RAFT-polymerization (proceeding in the presence of sulfur-containing compounds, which formula is $Z–C(=S)–S–R'$, where Z-stabilizing group, R'-outgoing group), valuable progress was obtained in the field of synthesis of new controlling agents (RAFT-agents), as well as in the field of research of kinetics and mathematical modeling; and for RAFT-polymerization in symmetrical RAFT-agents' presence, particularly, tritiocarbonates of formula $R'–S–C(=S)–S–R'$, it came to naught in practice: kinetics was studied in extremely general form [4] and mathematical modeling of process hasn't been carried out at all. Thus, the aim of this research is the kinetic modeling of polystyrene controlled radical polymerization initiated by 2,2'-asobis(isobutirnitrile) (AIBN), proceeding by reversible chain transfer mechanism and accompanied by "addition-fragmentation" in the presence of dibenzyltritiocarbonate (DBTC), and also the research of influence of the controlling factors (temperature, initial concentrations of monomer, AIBN and DBTC) on molecular-mass properties of polymer.

10.2   EXPERIMENTAL PART

Prior using of styrene (Aldrich, 99%), it was purified of aldehydes and inhibitors at triple cleaning in a separatory funnel by 10%-th (mass) solution of NaOH (styrene to solution ratio is 1:1), then it was scoured by distilled water to neutral reaction and after that it was dehumidified over $CaCl_2$ and rectified in vacuo.

AIBN (Aldrich, 99%) was purified of methanol by recrystallization.

DBTC was obtained by the method presented in research [4]. Masses of initial substances are the same as in [4]. Emission of DBTC was 81%. NMR ^{13}C (CCl$_3$D) δ, ppm: 41.37, 127.60, 128.52, 129.08, 134.75, and 222.35.

Examples of polymerization were obtained by dissolution of estimated quantity of AIBN and DBTC in monomer. Solutions were filled in tubes, 100 mm long, and having internal diameter of 3 mm, and after degassing in the mode of "freezing-defrosting" to residual pressure 0.01-mmHg column, the tubes were unsoldered. Polymerization was carried out at 60 °C.

Research of polymerization's kinetics was made with application of the calorimetric method on Calvet type differential automatic micro calorimeter DAK-1–1 in the mode of immediate record of heat emission rate in isothermal conditions at 60 °C. Kinetic parameters of polymerization were calculated basing on the calorimetric data as in the Ref. [5]. The value of polymerization enthalpy $\Delta H = -73.8$ kJ×mol^{-1} [5] was applied in processing of the data in the calculations.

Molecular-mass properties of polymeric samples were determined by gel-penetrating chromatography in tetrahydrofuran at 35°C on chromatograph GPCV 2000 "Waters." Dissection was performed on two successive banisters PLgel MIXED-C 300*7.5 mm, filled by stir gel with 5 μm vesicles. Elution rate-0.1 mL×min^{-1}. Chromatograms were processed in program "Empower Pro" with use of calibration by polystyrene standards.

10.3 MATHEMATICAL MODELLING OF POLYMERIZATION PROCESS

Kinetic scheme, introduced for description of styrene controlled radical polymerization process in the presence of trithio carbonates, includes the following phases.

1 Real initiation

$$I \xrightarrow{k_d} 2R(0) .$$

2. Thermal initiation [6]. It should be noted that polymer participation in thermal initiation reactions must reduce the influence thereof on molecular-mass distribution (MMD). However, since final mechanism of these reactions has not been ascertained in recording of balance differential equations for polymeric products so far, we will ignore this fact.

$$3M \xrightarrow{k_{i1}} 2R(1) ,$$

$$2M+P \xrightarrow{k_{i2}} R(1)+R(i) ,$$

$$2P \xrightarrow{\;k_{i3}\;} 2R(i).$$

In these three reactions summary concentration of polymer is recorded as P.

3. Chain growth

$$R(0)+M \xrightarrow{\;k_p\;} R(1),$$

$$R'+M \xrightarrow{\;k_p\;} R(1),$$

$$R(i)+M \xrightarrow{\;k_p\;} R(i+1).$$

4. Chain transfer to monomer

$$R(i)+M \xrightarrow{\;k_{tr}\;} P(i, 0, 0, 0) + R(1)$$

5. Reversible chain transfer [4]. As a broadly used assumption lately, we shall take that intermediates fragmentation rate constant doesn't depend on leaving radical's length [7].

$$R(i)+RAFT(0, 0) \underset{k_f}{\overset{k_{a1}}{\rightleftarrows}} Int(i, 0, 0) \underset{k_{a2}}{\overset{k_f}{\rightleftarrows}} RAFT(i, 0)+R' \qquad (I)$$

$$R(j)+RAFT(i, 0) \underset{k_f}{\overset{k_{a2}}{\rightleftarrows}} Int(i, j, 0) \underset{k_{a2}}{\overset{k_f}{\rightleftarrows}} RAFT(i, j)+R' \qquad (II)$$

$$R(k)+RAFT(i, j) \underset{k_f}{\overset{k_{a2}}{\rightleftarrows}} Int(i, j, k) \qquad (III)$$

6. Chain termination [4]. For styrene's RAFT-polymerization in the trithiocarbonates presence, besides reactions of radicals quadratic termination

$$R(0)+R(0) \xrightarrow{\;k_{t1}\;} R(0)\text{-}R(0),$$

$$R(0)+R' \xrightarrow{\;k_{t1}\;} R(0)\text{-}R',$$

$$R'+R' \xrightarrow{k_{t1}} R'\text{-}R',$$

$$R(0)+R(i) \xrightarrow{k_{t1}} P(i, 0, 0, 0),$$

$$R'+R(i) \xrightarrow{k_{t1}} P(i, 0, 0, 0),$$

$$R(j)+R(i\text{-}j) \xrightarrow{k_{t1}} P(i, 0, 0, 0)$$

are character reactions of radicals and intermediates cross termination.

$$R(0)+Int(i, 0, 0) \xrightarrow{k_{t2}} P(i, 0, 0, 0),$$

$$R(0)+Int(i, j, 0) \xrightarrow{k_{t2}} P(i, j, 0, 0),$$

$$R(0)+Int(i, j, k) \xrightarrow{k_{t2}} P(i, j, k, 0),$$

$$R'+Int(i, 0, 0) \xrightarrow{k_{t2}} P(i, 0, 0, 0),$$

$$R'+Int(i, j, 0) \xrightarrow{k_{t2}} P(i, j, 0, 0),$$

$$R'+Int(i, j, k) \xrightarrow{k_{t2}} P(i, j, k, 0),$$

$$R(j)+Int(i, 0, 0) \xrightarrow{k_{t2}} P(i, j, 0, 0),$$

$$R(k)+Int(i, j, 0) \xrightarrow{k_{t2}} P(i, j, k, 0),$$

$$R(m)+Int(i, j, k) \xrightarrow{k_{t2}} P(i, j, k, m).$$

In the introduced kinetic scheme: I, R(0), R(i), R', M, RAFT(i, j), Int(i, j, k), P(i, j, k, m) – reaction system's components (refer to Table 10.1); i, j, k, m–a number of monomer links in the chain; kd–a real rate constant of the initiation reaction; ki1, ki2, ki3, –thermal rate constants of the initiation reaction's; kp, ktr, ka1, ka2, kf, kt1, kt2 are the values of chain growth, chain transfer to monomer, radicals addition to low-molecular RAFT-agent, radicals addition to macromolecular RAFT-agent, intermediates fragmentation, radicals quadratic termination and radicals and inter-mediates cross termination reaction rate constants, respectively.

TABLE 10.1   Signs of Components in a Kinetic Scheme

I		Int(i, 0, 0)	
R(0)		Int(i, j, 0)	
R'		Int(i, j, k)	
M		P(i, 0, 0, 0)	
R(i)		P(i, j, 0, 0)	

RAFT(0,0)   P(i, j, k, 0)

$R(k)S$ ⎯ $SR(i)$
$SR(j)$,

RAFT(i, 0)   P(i, j, k, m)

RAFT(i, j)

The differential equations system describing this kinetic scheme, is as follows:

$$d[I]\,/\,dt = -k_d[I];$$

$$d[R(0)]/dt = 2f\,k_d[I] - [R(0)](k_p[M] + k_{t1}(2[R(0)] + [R'] + [R]) + k_{t2}(\sum_{i=1}^{\infty}[Int(i,\,0,\,0)] +$$

$$+\sum_{i=1}^{\infty}\sum_{j=1}^{\infty}[Int(i,\,j,\,0)] + \sum_{i=1}^{\infty}\sum_{j=1}^{\infty}\sum_{k=1}^{\infty}[Int(i,\,j,\,k)]));$$

$$d[M]\,/\,dt = -(k_p([R(0)] + [R'] + [R]) + k_{tr}[R])[M] - 3k_{i1}[M]^3 - 2k_{i2}[M]^2([M]_0 - [M]);$$

$$d[R']/dt = -k_p[R'][M] + 2k_f\sum_{i=1}^{\infty}[Int(i,\,0,\,0)] - k_{a2}[R']\sum_{i=1}^{\infty}[RAFT(i,\,0)] +$$

$$+k_f \sum_{i=1}^{\infty} \sum_{j=1}^{\infty} [Int(i, j, 0)] - k_{a2}[R'] \sum_{i=1}^{\infty} \sum_{j=1}^{\infty} [RAFT(i, j)] - [R'](k_{t1}([R(0)] + 2[R'] + [R]) +$$

$$+k_{t2}(\sum_{i=1}^{\infty} [Int(i, 0, 0)] + \sum_{i=1}^{\infty} \sum_{j=1}^{\infty} [Int(i, j, 0)] + \sum_{i=1}^{\infty} \sum_{j=1}^{\infty} \sum_{k=1}^{\infty} [Int(i, j, k)]));$$

$$d[RAFT(0,0)] / dt = -k_{a1}[RAFT(0,0)][R] + k_f \sum_{i=1}^{\infty} [Int(i, 0, 0)];$$

$$d[R(1)]/dt = 2k_{i1}[M]^3 + 2k_{i2}[M]^2([M]_0 - [M]) + 2k_{i3}([M]_0 - [M])^3 + k_p[M]([R(0)] + [R'] -$$

$$-[R(1)]) + k_{tr}[R(i)][M] - k_{a1}[R(1)][RAFT(0,0)] + k_f[Int(1, 0, 0)] -$$

$$-k_{a2}[R(1)] \sum_{i=1}^{\infty} [RAFT(i, 0)] + 2k_f[Int(1, 1, 0)] - k_{a2}[R(1)] \sum_{i=1}^{\infty} \sum_{j=1}^{\infty} [RAFT(i, j)] +$$

$$+3k_f[Int(1, 1, 1)] - [R(1)](k_{t1}([R(0)] + [R'] + [R]) + k_{t2}(\sum_{i=1}^{\infty} [Int(i, 0, 0)] +$$

$$+\sum_{i=1}^{\infty} \sum_{j=1}^{\infty} [Int(i, j, 0)] + \sum_{i=1}^{\infty} \sum_{j=1}^{\infty} \sum_{k=1}^{\infty} [Int(i, j, k)])), i = 2,...;$$

$$d[R(i)]/dt = k_p[M]([R(i-1)] - [R(i)]) - k_{tr}[R(i)][M] - k_{a1}[R(i)][RAFT(0,0)] + k_f[Int(i, 0, 0)] -$$

$$-k_{a2}[R(i)] \sum_{i=1}^{\infty} [RAFT(i, 0)] + 2k_f[Int(i, j, 0)] - k_{a2}[R(i)] \sum_{i=1}^{\infty} \sum_{j=1}^{\infty} [RAFT(i, j)] + 3k_f[Int(i, j, k)] -$$

$$-[R(i)](k_{t1}([R(0)] + [R'] + [R]) + k_{t2}(\sum_{i=1}^{\infty} [Int(i, 0, 0)] + \sum_{i=1}^{\infty} \sum_{j=1}^{\infty} [Int(i, j, 0)] +$$

$$+\sum_{i=1}^{\infty} \sum_{j=1}^{\infty} \sum_{k=1}^{\infty} [Int(i, j, k)])), \ i = 2,...;$$

$$d[Int(i, 0, 0)]/dt = k_{a1}[RAFT(0,0)][R(i)] - 3k_f[Int(i, 0, 0)] + k_{a2}[R'][RAFT(i, 0)] -$$

$$-k_{t2}[Int(i, 0, 0)]([R(0)] + [R'] + [R]);$$

$$d[Int(i, j, 0)]/dt = k_{a2}[RAFT(i, 0)][R(j)] - 3k_f[Int(i, j, 0)] + k_{a2}[R'][RAFT(i, j)] -$$

$-k_{t2}[\text{Int}(i, j, 0)]([R(0)]+[R']+[R]);$

$d[\text{Int}(i, j, k)]/dt=k_{a2}[\text{RAFT}(i, j)][R(k)]-3k_f[\text{Int}(i, j, k)]-k_{t2}[\text{Int}(i, j, k)]([R(0)]+[R']+[R]);$

$d[\text{RAFT}(i, 0)]/dt=2k_f[\text{Int}(i, 0, 0)]-k_{a2}[R'][\text{RAFT}(i, 0)]-k_{a2}[\text{RAFT}(i, 0)][R]+2k_f[\text{Int}(i, j, 0)];$

$d[\text{RAFT}(i, j)]/dt=k_f[\text{Int}(i, j, 0)]-k_{a2}[R'][\text{RAFT}(i, j)]-k_{a2}[\text{RAFT}(i, j)][R] + 3k_f[\text{Int}(i, j, k)];$

$$d[P(i, 0, 0, 0)] / dt=[R(i)](k_{t1}([R(0)]+[R'])+k_{tr}[M])+\frac{k_{t1}}{2}\sum_{j=1}^{i-1}[R(j)][R(i-j)] +$$

$+k_{t2}[\text{Int}(i, 0, 0)]([R(0)]+[R']);$

$$d[P(i, j, 0, 0)]/dt = k_{t2}([\text{Int}(i, j, 0)]([R(0)]+[R'])+\sum_{i+j=2}^{\infty}[R(j)][\text{Int}(i, 0, 0)]);$$

$$d[P(i, j, k, 0)]/dt = k_{t2}([\text{Int}(i, j, k)]([R(0)]+[R'])+\sum_{i+j+k=3}^{\infty}[R(k)][\text{Int}(i, j, 0)]);$$

$$d[P(i, j, k, m)]/dt = k_{t2}\sum_{i+j+k+m=4}^{\infty}[R(m)][\text{Int}(i, j, k)].$$

Here f – initiator's efficiency; $[R]=\sum_{i=1}^{\infty}[R(i)]$ – summary concentration of macroradicals; t – time.

A method of generating functions was used for transition from this equation system to the equation system related to the unknown MMD moments [8].

Number-average molecular mass (Mn), polydispersity index (PD) and weight-average molecular mass (Mw) are linked to MMD moments by the following expressions:

$$M_n = (\Sigma\mu_1 / \Sigma\mu_0)M_{ST}, \quad PD = \Sigma\mu_2\Sigma\mu_0 / (\Sigma\mu_1)^2, \quad M_w = PD \cdot M_n,$$

where $\Sigma\mu_0$, $\Sigma\mu_1$, $\Sigma\mu_2$ – sums of all zero, first and second MMD moments; $M_{ST} = 104$ g/mol – styrene's molecular mass.

10.4   RATE CONSTANTS

10.4.1   REAL AND THERMAL INITIATION

The efficiency of initiation and temperature dependence of polymerization real initiation reaction rate constant by AIBN initiator are determined basing on the data in this research, which have established a good reputation for mathematical modeling of leaving in mass styrene radical polymerization [6]:

$$f = 0.5, \ k_d = 1.58 \cdot 10^{15} e^{-15501/T}, \ s^{-1},$$

where T–temperature, K.

As it was established in the research, thermal initiation reactions' rates constants depend on the chain growth reactions rate constants, the radicals quadratic termination and the monomer initial concentration:

$$k_{i1} = 1.95 \cdot 10^{13} \frac{k_{t1}}{k_p^2 M_0^3} e^{-20293/T}, \ L^2 \cdot mol^{-2} \cdot s^{-1};$$

$$k_{i2} = 4.30 \cdot 10^{17} \frac{k_{t1}}{k_p^2 M_0^3} e^{-23878/T}, \ L^2 \cdot mol^{-2} \cdot s^{-1};$$

$$k_{i3} = 1.02 \cdot 10^8 \frac{k_{t1}}{k_p^2 M_0^2} e^{-14807/T}, \ L \cdot mol^{-1} \cdot s^{-1}. \ [6].$$

10.4.2   CHAIN TRANSFER TO MONOMER REACTION'S RATE CONSTANT

On the basis of the data in research [6]:

$$k_{tr} = 2.31 \cdot 10^6 e^{-6376/T}, \ L \cdot mol^{-1} \cdot s^{-1}.$$

10.4.3   RATE CONSTANTS FOR THE ADDITION OF RADICALS TO LOW-MOLECULAR AND MACROMOLECULAR RAFT-AGENTS

In research [9], it was shown by the example of dithiobenzoates at first that chain transfer to low- and macromolecular RAFT-agents of rate constants are functions of respective elementary constants. Let us demonstrate this for our process. For this record, the change of concentrations [Int(i, 0, 0)], [Int(i, j, 0)], [RAFT(0,0)] and [RAFT (i, 0)] in quasistationary approximation for the initial phase of polymerization is as follows:

$$d[Int(i, 0, 0)]/dt = k_{a1}[RAFT(0,0)][R] - 3k_f[Int(i, 0, 0)] \approx 0, \tag{1}$$

$$d[Int(i, j, 0)]/dt = k_{a2}[RAFT(i, 0)][R] - 3k_f[Int(i, j, 0)] \approx 0, \qquad (2)$$

$$d[RAFT(0,0)] / dt = -k_{a1}[RAFT(0,0)][R] + k_f[Int(i, 0, 0)], \qquad (3)$$

$$d[RAFT(i, 0)]/dt = 2k_f[Int(i, 0, 0)] - k_{a2}[RAFT(i, 0)][R] + 2k_f[Int(i, j, 0)]. \qquad (4)$$

The Eq. (1) expresses the following concentration $[Int(i, 0, 0)]$:

$$[Int(i, 0, 0)] = \frac{k_{a1}}{3k_f}[RAFT(0,0)][R].$$

Substituting the expansion gives the following $[Int(i, 0, 0)]$ expression to Eq. (3):

$$d[RAFT(0,0)] / dt = -k_{a1}[RAFT(0,0)][R] + k_f \frac{k_{a1}}{3k_f}[RAFT(0,0)][R].$$

After transformation of the last equation, we have:

$$\frac{d[RAFT(0,0)]}{[RAFT(0,0)]} = -\frac{2}{3}k_{a1}[R]dt.$$

Solving this equation (initial conditions): $t = 0$, $[R] = [R]_0 = 0$, $[RAFT(0,0)] = [RAFT(0,0)]_0$), we obtain:

$$\ln\frac{[RAFT(0,0)]}{[RAFT(0,0)]_0} = -\frac{2}{3}k_{a1}[R]t. \qquad (5)$$

To transfer from time t, being a part of Eq. (5), to conversion of monomer C_M, we put down a balance differential equation for monomer concentration, assuming that at the initial phase of polymerization, thermal initiation and chain transfer to monomer are not of importance:

$$d[M] / dt = -k_p[R][M] \qquad (6)$$

Transforming the Eq. (6) with its consequent solution at initial conditions $t = 0$, $[R] = [R]_0 = 0$, $[M] = [M]_0$:

$$d[M] / [M] = -k_p[R]dt,$$

$$\ln \frac{[M]}{[M]_0} = -k_p[R]t \cdot \qquad (7)$$

Link rate $[M]/[M]_0$ with monomer conversion (C_M) in an obvious form like this:

$$C_M = \frac{[M]_0 - [M]}{[M]_0} = 1 - \frac{[M]}{[M]_0},$$

$$\frac{[M]}{[M]_0} = 1 - C_M.$$

We substitute the last ratio to Eq. (7) and express time t:

$$t = \frac{-\ln(1 - C_M)}{k_p[R]}. \qquad (8)$$

After substitution of the Eq. (8) by the Eq. (5), we obtain the next equation:

$$\ln \frac{[RAFT(0,0)]}{[RAFT(0,0)]_0} = \frac{2}{3} \frac{k_{al}}{k_p} \ln(1 - C_M). \qquad (9)$$

By analogy with introduced $[M]/[M]_0$ to monomer conversion, reduce ratio $[RAFT(0,0)] / [RAFT(0,0)]_0$ to conversion of low-molecular RAFT-agent – $C_{RAFT(0,0)}$. As a result, we obtain:

$$\frac{[RAFT(0,0)]}{[RAFT(0,0)]_0} = 1 - C_{RAFT(0,0)}. \qquad (10)$$

Substitute the derived expression for $[RAFT(0,0)] / [RAFT(0,0)]_0$ from Eq. (10) to Eq. (9):

$$\ln(1 - C_{RAFT(0,0)}) = \frac{2}{3} \frac{k_{al}}{k_p} \ln(1 - C_M). \qquad (11)$$

In the Ref. [9], the next dependence of chain transfer to low-molecular RAFT-agent constant C_{trl} is obtained on the monomer and low-molecular RAFT-agent conversions:

$$C_{trl} = \frac{\ln(1 - C_{RAFT(0,0)})}{\ln(1 - C_M)}. \qquad (12)$$

Comparing Eqs. (12) and (11), we obtain dependence of chain transfer to low-molecular RAFT-agent constant C_{tr1} on the constant of radical's addition to macro-molecular RAFT-agent and chain growth reaction rate constant:

$$C_{tr1} = \frac{2}{3} \frac{k_{a1}}{k_p}. \tag{13}$$

From Eq. (13), we derive an expression for constant k_{a1}, which will be based on the following calculation:

$$k_{a1} = 1.5 C_{tr1} k_p, \; L \cdot mol^{-1} \cdot s^{-1}$$

As a numerical value for C_{tr1}, we assume value 53, derived in research [4] on the base of Eq. (12), at immediate experimental measurement of monomer and low-molecular RAFT-agent conversions. Since chain transfer reaction in RAFT-polymerization is usually characterized by low value of activation energy, compared to activation energy of chain growth, it is supposed that constant C_{tr1} doesn't depend or slightly depends on temperature. We will propose as an assumption that C_{tr1} doesn't depend on temperature [10].

By analogy with k_{a1}, we deduce equation for constant k_{a2}. From Eq. (2) we express such concentration $[Int(i, j, 0)]$:

$$[Int(i, j, 0)] = \frac{k_{a2}}{3k_f} [RAFT(i, 0)][R].$$

Substitute expressions, derived for $[Int(i, 0, 0)]$ and $[Int(i, j, 0)]$ in Eq. (4):

$$d[RAFT(i, 0)]/dt = \frac{2}{3} k_{a1}[RAFT(0,0)][R] - \frac{1}{3} k_{a2}[RAFT(i, 0)][R]. \tag{14}$$

Since in the end it was found that constant of chain transfer to low-molecular RAFT-agent C_{tr1} is equal to divided to constant k_p coefficient before expression $[RAFT(0,0)][R]$ in the balance differential equation for $[RAFT(0,0)]$, from Eq. (14) for constant of chain transfer to macromolecular RAFT-agent, we obtain the next expression:

$$C_{tr2} = \frac{1}{3} \frac{k_{a2}}{k_p}.$$

From the last equation we obtain an expression for constant k_{a2}, which based on the following calculation:

$$k_{a2} = 3 C_{tr2} k_p, \; L \cdot mol^{-1} \cdot s^{-1} \tag{15}$$

In research [4] on the base of styrene and DBTK, macromolecular RAFT-agent was synthesized, thereafter with a view to experimentally determine constant C_{tr2}, polymerization of styrene was performed with the use of the latter. In the course of experiment, it may be supposed that constant C_{tr2} depends on monomer and macromolecular RAFT-agent conversions by analogy with Eq. (12). As a result directly from the experimentally measured monomer and macromolecular RAFT-agent conversions, value C_{tr2} was derived, equal to 1000. On the ground of the same considerations as for that of C_{tr1}, we assume independence of constant C_{tr2} on temperature.

10.4.4 RATE CONSTANTS OF INTERMEDIATES FRAGMENTATION, TERMINATION BETWEEN RADICALS AND TERMINATION BETWEEN RADICALS AND INTERMEDIATES

In research [4] it was shown, that RAFT-polymerization rate is determined by this equation:

$$\left(W_0 / W\right)^2 = 1 + \frac{k_{t2}}{k_{t1}} K[RAFT(0,0)]_0 + \frac{k_{t3}}{k_{t1}} K^2 [RAFT(0,0)]_0^2,$$

where W_0 and W-polymerization rate in the absence and presence of RAFT-agent, respectively, s^{-1}; K constant of equilibrium (III), $L \cdot mol^{-1}$; k_{t3} constant of termination between two intermediates reaction rate, $L \cdot mol^{-1} \cdot s^{-1}$ [11].

For initiated AIBN styrene polymerization in DBTC's presence at 80°C, it was shown that intermediates quadratic termination wouldn't be implemented and RAFT-polymerization rate was determined by equation [4]:

$$\left(W_0 / W\right)^2 = 1 + 8[RAFT(0,0)]_0.$$

Since $\frac{k_{t2}}{k_{t1}} \approx 1$, then at 80°C $K = 8 \, L \cdot mol^{-1}$ [4]. In order to find dependence of constant K on temperature, we made research of polymerization kinetics at 60°C. It was found, (Fig. 10.1) that the results of kinetic measurements well rectify in coordinates $\left(W_0 / W\right)^2 = f([RAFT(0,0)]_0)$. At 60°C, $K = 345$ $L \cdot mol^{-1}$ was obtained. Finally dependence of equilibrium constant on temperature has been determined in the form of Vant-Goff's equation:

$$K = 4.85 \cdot 10^{-27} e^{22123/T}, \, L \cdot mol^{-1}. \tag{16}$$

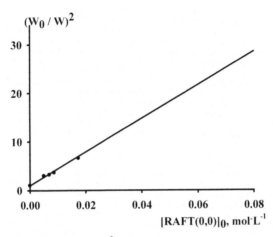

FIGURE 10.1   Dependence $(W_0 / W)^2$ on DBTC concentration at 60°C.

In compliance with the equilibrium (III), the constant is equal to $K=\dfrac{k_{a2}}{3k_f}$, L·mol⁻¹.

Hence, reactions of intermediates fragmentation rate constant will be as such:

$$k_f=\frac{k_{a2}}{3K}, \; s^{-1} \tag{17}$$

The reactions of intermediates fragmentation rate constant were built into the model in the form of dependence (the Eq. (17)) considering Eqs. (15) and (16).

As it has been noted above, ratio k_{t2} / k_{t1} equals approximately to one, therefore it will taken, that $k_{t2} \approx k_{t1}$ [4]. For description of gel-effect, dependence as a function of monomer conversion C_M and temperature T (K) [12] was applied:

$$k_{t2} \approx k_{t1} \approx 1.255 \cdot 10^9 e^{-844/T} e^{-2(A_1 C_M + A_2 C_M^2 + A_3 C_M^3)}, \; L \cdot mol^{-1} \cdot s^{-1},$$

where $A_1=2.57-5.05 \cdot 10^{-3}T$; $A_2=9.56-1.76 \cdot 10^{-2}T$; $A_3=-3.03+7.85 \cdot 10^{-3}T$.

10.4.5   RATE CONSTANT FOR CHAIN GROWTH

The method of polymerization, being initiated by pulse laser radiation [13] is used for determination of rate constant for chain growth k_p lately. It is anticipated that such an estimation method is more correct, than the traditionally used revolving sector method [12]. We made our choice on temperature dependence of the rate constant for chain growth that was derived on the ground of method of polymerization, being initiated by pulse laser radiation:

$$k_p = 4.27 \cdot 10^7 e^{-3910/T} \text{ , L·mol}^{-1}\text{·s}^{-1}, \qquad (18)$$

Since this dependence is more adequately describes the change of polymerization reduced rate with monomer conversion in the network of the developed mathematical model (Fig. 10.2), than temperature dependence, which is derived by revolving sector method [12]:

$$k_p = 1.057 \cdot 10^7 e^{-3667/T} \text{ , L·mol}^{-1}\text{·s}^{-1}. \qquad (19)$$

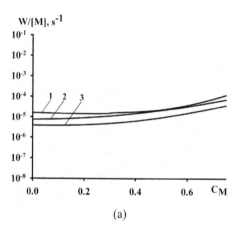

(a)

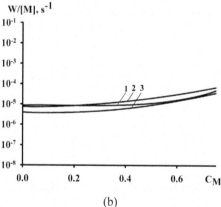

(b)

FIGURE 10.2   Dependence of initiated AIBN ($[I]_0$=0.01 mol·L^{-1}) styrene polymerization reduced rate on monomer conversion at 60°C (1. experiment; 2. estimation by introduced in this research mathematical model with temperature dependence of k_p (see Eq. (18)); 3. estimation by introduced in this research mathematical model with temperature dependence of k_p (see Eq. (19)): $[\text{RAFT}(0,0)]_0 = 0$ mol·L^{-1} (a), 0.007 (b)).

10.5 MODEL'S ADEQUACY

The results of polystyrene molecular-mass properties calculations by the introduced mathematical model are presented in Figs. 10.3 and 10.4. Mathematical model of styrene RAFT-polymerization in the presence of trithiocarbonates, taking into account the radicals and intermediates cross termination, adequately describes the experimental data that prove the process mechanism, built in the model. The essential proof of the mechanism correctness is that in case of conceding the absence of radicals and intermediates cross termination the experimental data wouldn't substantiate theoretical calculation by the mathematical model, introduced in this assumption (Fig. 10.5).

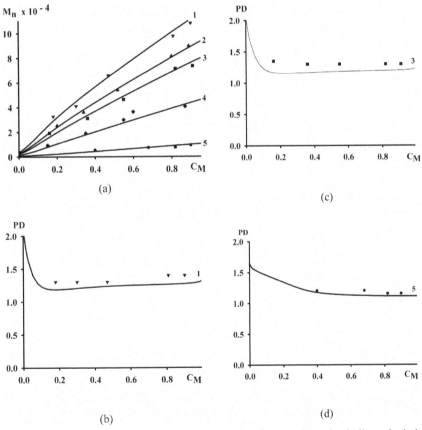

FIGURE 10.3 Dependence of number-average molecular mass (a) and polydispersity index (b)–(d) on monomer conversion for being initiated by AIBN ($[I]_0 = 0.01$ mol·L^{-1}) styrene bulk RAFT-polymerization at 60°C in the presence of DBTC (lines estimation by model; points experiment): $[RAFT(0,0)]_0 = 0.005$ mol·L^{-1} (1), 0.007 (2), 0.0087 (3), 0.0174 (4), 0.087 (5).

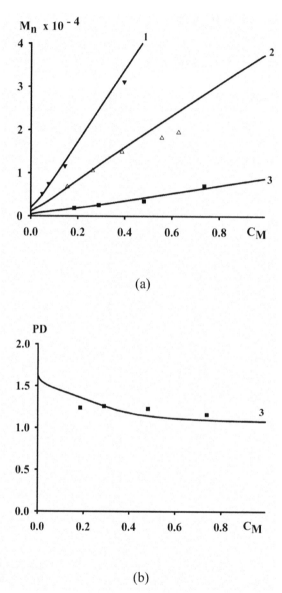

(a)

(b)

FIGURE 10.4   Dependence of number-average molecular mass (a) and polydispersity index (b) on monomer conversion for being initiated by AIBN ($[I]_0$=0.01 mol·L⁻¹) styrene bulk RAFT-polymerization at 80°C in DBTC presence (lines estimation by model; points experiment): $[RAFT(0,0)]_0 = 0.01$ mol·L⁻¹ (1), 0.02 (2), 0.1 (3) [4].

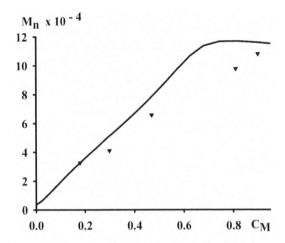

FIGURE 10.5 Dependence of number-average molecular mass on monomer conversion for initiated AIBN ($[I]_0$=0.01 mol·L^{-1}) styrene bulk RAFT-polymerization at 60°C in DBTC presence $[RAFT(0,0)]0 = 0.005$ mol·L^{-1} (lines estimation by model assuming that radicals and intermediates cross termination are absent; points experiment).

Due to adequacy of the model realization at numerical experiment it became possible to determine the influence of process controlling factors on polystyrene molecular-mass properties.

10.6   NUMERICAL EXPERIMENT

Research of influence of the process controlling factors on molecular-mass properties of polystyrene, synthesized by RAFT-polymerization method in the presence of AIBN and DBTC, was made in the range of initial concentrations of: initiator 0–0.1 mol·L^{-1}, monomer 4.35–8.7 mol·L^{-1}, DBTC 0.001–0.1 mol·L^{-1}; and at temperatures 60–120°C.

10.6.1   THE INFLUENCE OF AIBN INITIAL CONCENTRATION BY NUMERICAL EXPERIMENT

It was set forth that generally in some other conditions, with increase of AIBN initial concentration number-average, the molecular mass of polystyrene decreases (Fig. 10.6). At all used RAFT-agent initial concentrations, there is a linear or close to linear growth of number-average molecular mass of polystyrene with monomer conversion. This means that even the lowest RAFT-agent initial concentrations affect the process of radical polymerization. It should be noted that at high RAFT-agent initial concentrations (Fig. 10.7) the change of AIBN initial concentration practically doesn't have any influence on number-average molecular mass of polystyrene. But at increased temperatures (Fig. 10.8), in case of high AIBN initial concentration, it is comparable to high RAFT-agent initial concentration; polystyrene molecular mass would be slightly decreased due to thermal initiation.

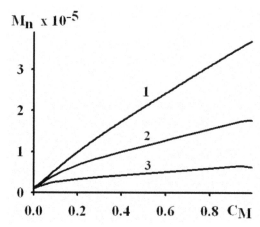

FIGURE 10.6   Dependence of number-average molecular mass M_n on monomer conversion C_M (60°C) $[M]_0 = 6.1$ mol·L⁻¹, $[RAFT(0, 0)]_0 = 0.001$ mol·L⁻¹, $[I]_0 = 0.001$ mol·L⁻¹ (1), 0.01 (2), 0.1 (3).

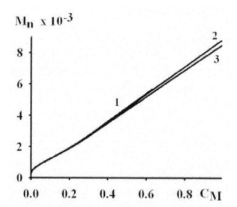

FIGURE 10.7   Dependence of number-average molecular mass M_n on monomer conversion C_M (60°C) $[M]_0 = 8.7$ mol·L⁻¹, $[RAFT(0, 0)]_0 = 0.1$ mol·L⁻¹, $[I]_0 = 0.001$ mol·L⁻¹ (1), 0.01 (2), 0.1 (3).

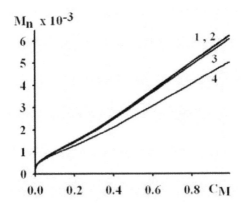

FIGURE 10.8   Dependence of number-average molecular mass M_n on monomer conversion C_M (120°C) $[M]_0 = 6.1$ mol·L^{-1}, $[RAFT(0, 0)]_0 = 0.1$ mol·L^{-1}, $[I]_0 =$ mol·L^{-1} (1), 0.001 (2), 0.01 (3), 0.1 (4).

Since the main product of styrene RAFT-polymerization process, proceeding in the presence of trithiocarbonates, is a narrow-dispersed high-molecular RAFT-agent (marked in kinetic scheme as RAFT(i, j)), which is formed as a result of reversible chain transfer, and widely dispersed (minimal polydispersity 1.5) polymer, forming by the radicals quadratic termination, so common polydispersity index of synthesizing product is their ratio. In a broad sense, with increase of AIBN initial concentration, the part of widely dispersed polymer, which is formed as a result of the radicals quadratic termination, increase in mixture, thereafter general polydispersity index of synthesizing product increases.

However, at high temperatures this regularity can be discontinued at low RAFT-agent initial concentrations the increase of AIBN initial concentration leads to a decrease of polydispersity index (Fig. 10.9, curves 3 and 4). This can be related only thereto that at high temperatures thermal initiation and elementary reactions rate constants play an important role, depending on temperature, chain growth and radicals quadratic termination reaction rate constants, monomer initial concentration in a complicated way [6]. Such complicated dependence makes it difficult to analyze the influence of thermal initiation role in process kinetics; therefore the expected width of MMD of polymer, which is expected to be synthesized at high temperatures, can be estimated in every specific case in the frame of the developed theoretical regularities.

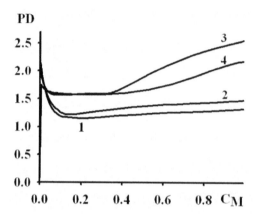

FIGURE 10.9   Dependence of polydispersity index PD on monomer conversion C_M (120°C) $[M]_0 = 8.7$ mol·L⁻¹, $[RAFT(0, 0)]_0 = 0.001$ mol·L⁻¹, $[I]_0 = 0$ mol·L⁻¹ (1), 0.001 (2), 0.01 (3), 0.1 (4).

Special attention shall be drawn to the fact that for practical objectives, realization of RAFT-polymerization process without an initiator is of great concern. In all cases at high temperatures as the result of styrene RAFT-polymerization implementation in the presence of RAFT-agent without AIBN, more high-molecular (Fig. 10.10) and more narrow-dispersed polymer (Fig. 10.9, curve 1) is built-up than in the presence of AIBN (Fig. 10.9, curves 2–4).

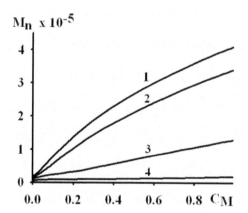

FIGURE 10.10   Dependence of number-average molecular mass M_n on monomer conversion C_M (120°C) $[M]_0 = 8.7$ mol·L⁻¹, $[RAFT(0, 0)]_0 = 0.001$ mol·L⁻¹, $[I]_0 = 0$ mol·L⁻¹ (1), 0.001 (2), 0.01 (3), 0.1 (4).

10.6.2 THE INFLUENCE OF MONOMER INITIAL CONCENTRATION BY NUMERICAL EXPERIMENT

In other identical conditions, the decrease of monomer initial concentration reduces the number-average molecular mass of polymer. Polydispersity index doesn't practically depend on monomer initial concentration.

10.6.3 INFLUENCE OF RAFT-AGENT INITIAL CONCENTRATION BY NUMERICAL EXPERIMENT

In other identical conditions, increase of RAFT-agent initial concentration reduces the number-average molecular mass and polydispersity index of polymer (Fig. 10.11).

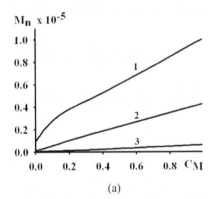

(a)

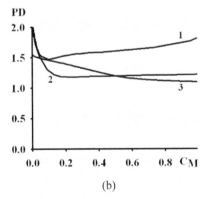

(b)

FIGURE 10.11 Dependence of number-average molecular mass M_n (a) and polydispersity index PD (b) on monomer conversion C_M (90°C) $[I]_0 = 0.01$ mol·L^{-1}, $[M]_0 = 6.1$ mol·L^{-1}, $[RAFT(0, 0)]_0 = 0.001$ mol·L^{-1} (1), 0.01 (2), 0.1 (3).

10.6.4   THE INFLUENCE OF TEMPERATURE BY NUMERICAL EXPERIMENT

Generally, in other identical conditions, the increase of temperature leads to a decrease of number-average molecular mass of polystyrene (Fig. 10.12a). Thus, polydispersity index increases (Fig. 10.12b). If RAFT-agent initial concentration greatly exceeds AIBN initial concentration, then the temperature practically doesn't influence the molecular-mass properties of polystyrene.

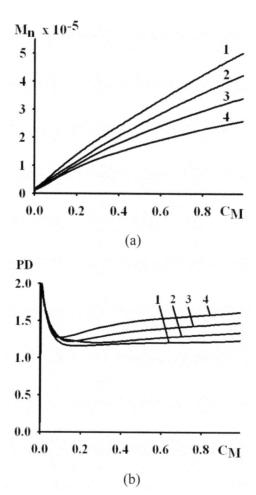

(a)

(b)

FIGURE 10.12   Dependence of number-average molecular mass M_n (a) and polydispersity index PD (b) on monomer conversion C_M $[I]_0 = 0.001$ mol·L^{-1}, $[M]_0 = 8.7$ mol·L^{-1}, [RAFT(0, 0)]$_0 = 0.001$ mol·L^{-1}, T = 60°C (1), 90 (2), 120 (3), 150 (4)

10.7   CONCLUSION

The kinetic model developed in this research allows an adequate description of molecular-mass properties of polystyrene, obtained by controlled radical polymerization, which proceeds by reversible chain transfer mechanism and accompanied by "addition-fragmentation." This means, that the model can be used for development of technological applications of styrene RAFT-polymerization in the presence of trithiocarbonates.

Researches were supported by Russian Foundation for Basic Research (project. no. 12–03–97050-r_povolzh'e_a).

KEYWORDS

- **Controlled radical polymerization**
- **Dibenzyltritiocarbonate**
- **Mathematical modeling**
- **Polystyrene**
- **Reversible addition-fragmentation chain transfer**

REFERENCES

1. Matyjaszewski, K. (2009). Controlled/Living Radical Polymerization: Progress in ATRP, D.C.: American Chemical Society, Washington.
2. Matyjaszewski, K. (2009). Controlled/Living Radical Polymerization: Progress in RAFT, DT, NMP and OMRP, D.C.: American Chemical Society, Washington.
3. Barner-Kowollik, C. (2008). Handbook of RAFT Polymerization. Wiley VCH Verlag GmbH, Weinheim.
4. Chernikova, E. V., Terpugova. P. S., Garina, E. S., & Golubev, V. B. (2007). Controlled radical polymerization of styrene and n-butyl acrylate mediated by tritiocarbonates. *Polymer Science, 49(2)*, 108.
5. Stephen. Z. D. (2002). Cheng: *Handbook of Thermal Analysis and Calorimetry 3, Applications to Polymers and Plastics*: New York, Elsevier.
6. Kuzub, L. I., Peregudov, N. I., & Irzhak, V. I. (2005). *Polymer Science, 47,* 1063.
7. Zetterlund, P. B., & Perrier S. (2011). *Macromolecules, 44,* 1340.
8. Biesenberger, J. A., & Sebastian, D. H. (1983). *Principles of Polymerization Engineering*. John Wiley & Sons Inc., New York.
9. Chong, Y. K., Krstina, J., Le, T. P. T., Moad, G., Postma, A., Rizzardo, E., & Thang, S. H. (2003). *Macromolecules, 36, 2256.*
10. Goto, A., Sato, K., Tsujii, Y., Fukuda, T., Moad, G., Rizzardo, E., & Thang, S. H. (2001). *Macromolecules, 34,* 402.

11. Kwak, Y., Goto, A. & Fukuda, T. (2004). *Macromolecules, 37,* 1219.
12. Hui, A. W., & Hamielec, A. E. (1972). *J Appl. Polym. Sci., 16,* 749.
13. Li, D., & Hutchinson, R. A. (2007). *Macromolecular Rapid Communications, 28,* 1213.

CHAPTER 11

A RESEARCH NOTE ON SEMIEMPIRICAL AND DFT MODELING OF THE IR SPECTRA OF BENZOYL PEROXIDE DERIVATIVES

N. A. TUROVSKIJ, YU. V. BERESTNEVA, E. V. RAKSHA,
E. N. PASTERNAK, I. A. OPEIDA, and G. E. ZAIKOV

CONTENTS

ABSTRACT

The infrared spectra of the benzoyl peroxide symmetrical derivatives (4-R-PhCOO)$_2$ with R: NO$_2$–, CF$_3$–, CF$_3$O–, I–, Br–, Cl–, F–, H–, CH$_3$–, CH$_3$O– were studied by the semiempirical methods. There is a linear relationship between the frequencies of the normal vibrations of the experimental and calculated (PM6, PDDG and AM1) spectra for this series of peroxides. The effect of the DFT level on the normal vibrations frequencies of C = O group of benzoyl peroxide was estimated. The best reproduction of these frequencies is observed in the case of BLYP calculation method with 6-311G (d, p) basis set.

11.1   AIMS AND BACKGROUND

The methods of the molecular spectroscopy rightfully occupy one of the leading places in the study of the peroxide compounds structure. The IR spectroscopy is successfully used to establish the structure of the synthesized peroxides as well as for their identification and kinetic studies [1]. The IR spectra play the role of indicator of the oxidative stability and peroxide value in the foods oxidative modification [1]. An important feature of the IR spectroscopy is the absence of damaging effects of the infrared light quanta on the peroxide molecule. This method allows one to investigate peroxides in any aggregate state. Analytical capabilities of the infrared spectroscopy in the study of processes involving diacyl peroxides are possible due to the presence in peroxides extremely characteristic frequency of the intensity ratio and "doublet" form in the 1750–1820 cm^{-1}, which corresponds to the absorption of carbonyl group of the diacyl peroxide fragment (-C (O)-O-O-C (O)-).

There are two absorption bands belonging to the v(C=O) in the spectra of symmetric diacyl peroxides that explained by the resonant interaction of carbonyl groups [2], as well as the in-phase (v$_s$) and antiphase (v$_{as}$) vibrations of the C=O groups [3]. The bands that belong to the group vibration absorption molecular fragments as a whole with the traditional interpretation of the assignment of bands in the IR spectrum were allocated for aliphatic peroxides – within 1132–1180 and 1045–1076 cm^{-1}, heterocyclic 1265 ± 10 and 1070 ± 10 cm^{-1} aromatic 1048–1074 cm^{-1}. The largest differences are observed in assigning of the absorption bands of the peroxide fragment -O-O-, which most researchers refer to the strip in the area of 850 ± 15 cm^{-1} [4].

The development of quantum chemistry and the growth of computing power have led to the fact that modern semiempirical, DFT and ab initio methods of the quantum chemistry can significantly improve the speed and accuracy of calculations of various physical and chemical characteristics of the objects or processes and in many cases allow to achieve precise agreement with the experimental data [5]. This makes it possible to predict the molecular force constants and frequencies of normal vibrations. The second derivative of the total energy of the molecule on the internal coordinates and on the electric field determines the intensity of the IR absorption [6]. Calculation of normal vibrations allows estimating the contributions of various fragments of the molecule in the distribution of potential energy at different frequencies. Comparison of experimental and calculated vibration spectra significantly simplifies the correct assignment of the bands to a certain type of vibrations. Calculated methodology successfully used to determine the characteristic frequencies of the peroxides with unknown individual IR spectra [7, 8].

The aim of this study is the assessment of opportunities of DFT and semiempirical methods for the reproduction and prediction of the diacyl peroxides IR spectra. This chapter presents the results of IR spectra molecular modeling of the benzoyl peroxide symmetrical derivatives (4-R-PhCOO)$_2$, R: NO$_2$–, CF$_3$–, CF$_3$O–, I–, Br–, Cl–, F–, H–, CH$_3$–, CH$_3$O– by the semiempirical (AM1, PM6, PDDG) methods. The IR spectra of the benzoyl peroxide were investigated also at the DFT level.

11.2 EXPERIMENTAL PART

Parameters of the molecular geometry, electronic structure and thermodynamic properties of the benzoyl peroxide (BPO) molecule and its symmetrical derivatives were calculated by the GAUSSIAN09 [9]. The molecular geometry optimization of all objects was carried out at the first stage of the work. The calculation of harmonic frequencies of vibrations and thermodynamic parameters were performed after that. The stationary points obtained after the molecular geometry optimization were identified as minima, as there were no negative values of analytic harmonic vibration frequencies for them. The reaction center of the peroxide compounds is a peroxide bond -O-O-. Therefore, selection criterion for the quantum chemical calculation method was the best reproduction of the peroxide moiety molecular geometry.

To solve the problem of choosing the optimal method for molecular geometry optimization and calculation of infrared spectra of BPO and its symmetrical derivatives the evaluation parameters of the structure of these compounds and the strength of peroxide bond were performed by semiempirical (AM1 [10], PM6 [11], PDDG [12]) and DFT methods (BLYP [13], B1LYP [14], B3LYP [12], B971 [15], B972 [16], BH and HLYP [17], M06HF [18], O3LYP [19], PBE1PBE [20], PBEh1PBE [21], X3LYP [22]). The solvent effect was taken into account in the CPCM approximation [23]. Visualization of the calculated IR spectra was carried out using Chemcraft 1.6 [24].

The experimental IR spectra of the benzoyl peroxide (Fig. 11.1) and its derivatives in CH$_2$Cl$_2$ used in the study were taken from [4]. The following symmetric diacyl peroxides: (4-NO$_2$-C$_6$H$_4$-COO)$_2$, (4-CF$_3$-C$_6$H$_4$-COO)$_2$, (4-CF$_3$O-C$_6$H$_4$-COO)$_2$, (4-I-C$_6$H$_4$-COO)$_2$, (4-Br-C$_6$H$_4$-COO)$_2$, (4-Cl-C$_6$H$_4$-COO)$_2$, (C$_6$H$_5$-COO)$_2$, (4-CH$_3$-C$_6$H$_4$-COO)$_2$, (4-CH$_3$-O-C$_6$H$_4$-COO)$_2$ have been investigated.

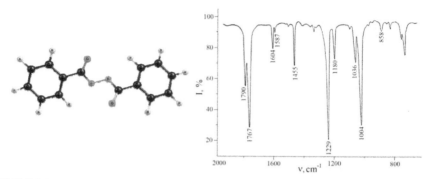

FIGURE 11.1   Structural model of the benzoyl peroxide (PM6 method) and its experimental IR spectrum in CH_2Cl_2, [ROOR] = 0.14 mol/dm³, thickness 0.058 mm (according to Ref. [4]).

11.3   RESULTS AND DISCUSSION

The equilibrium configuration of the benzoyl peroxide molecule was obtained by the PM6, AM1 and PDDG semiempirical methods using the CPCM solvation model (Table 11.1). The equilibrium configuration geometry of the benzoyl peroxide was used in the calculation of IR spectra.

Calculation results listed in Table 11.1 show the best reproduction of the C=O group normal vibrations frequencies in the case of PM6 calculation method application.

TABLE 11.1   Parameters of the Benzoyl Peroxide Molecular Geometry in CH_2Cl_2

Parameters	Experiment [4, 25, 26]	PM6/CPCM	AM1/CPCM	PDDG/CPCM
C=O, Å	1.190	1.205	1.230	1.216
O–O, Å	1.460	1.423	1.292	1.440
C–O, Å	1.380	1.408	1.406	1.367
O-O-C, °	111.0	109.8	114.5	113.0
C-O-O-C, °	91.0	101.5	91.5	105.9
μ, D	1.60	1.53	1.71	1.88
v_{as}(C=O), cm⁻¹	1767	1800	2065	1993
v_s(C=O), cm⁻¹	1790	1808	2076	2004

Liner relationships between the experimental (v_{exp}) and calculated (v_{calc}) values of the normal vibrations are observed for the BPO IR spectra (Fig. 11.2). These relationships are described by Eqs. (1)–(3) for PM6, PDDG and AM1 quantum chemical calculation methods respectively.

$$V_{exp} = (195 \pm 26) + (0.79 \pm 0.02) \cdot V_{calc}, n = 19, R = 0.995 \qquad (1)$$

$$V_{exp} = (119 \pm 29) + (0.82 \pm 0.02) \cdot V_{calc}, n = 19, R = 0.995 \qquad (2)$$

$$V_{exp} = (195 \pm 26) + (0.79 \pm 0.02) \cdot V_{calc}, n = 19, R = 0.995 \qquad (3)$$

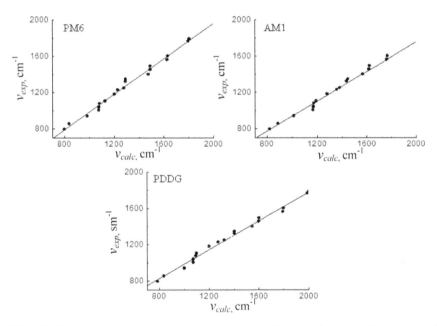

FIGURE 11.2 Relationships between the experimental (v_{exp}) and calculated (v_{calc}) normal vibrations frequencies of the benzoyl peroxide IR spectrum in CH_2Cl_2 (CPCM solvation model).

The influence of the structure of the benzoyl peroxide symmetrical derivatives (R-PhC (O)-O-O-C (O) Ph-R) on the position of the carbonyl groups absorption band in the IR spectra has been also investigated. Calculations were carried out in the approximation of AM1, PM6 and PDDG semiempirical methods. The best correspondence between the experimental and calculated parameters is observed in the case of PM6 method (Table 11.2). Introduction of electron-donor substituents into the structure of benzoyl peroxide leads to a shift of the normal vibrations mode of $C = O$ groups towards the field of short waves, while shift of the mode towards the region of long waves is observed in the case of electron- acceptor substituents in peroxide molecule.

TABLE 11.2   The Normal Vibrations Frequencies of the $C = O$ groups $(n(C = O),\ cm^{-1})$ of the BPO Derivatives R-PhC(O)-O-O-C(O)Ph-R

R-	Experiment [4]		PM6		AM1		PDDG	
	v_{as}	v_s	v_{as}	v_s	v_{as}	v_s	v_{as}	v_s
NO$_2$-	1780	1801	1851	1847	2101	2090	2029	2021
CF$_3$-	1776	1798	1849	1845	2100	2090	2027	2020
CF$_3$O-	1771	1795	1846	1843	2098	2087	2025	2018
I-	1771	1792	1846	1842	2099	2088	2024	2017
Br-	1772	1792	1846	1842	2099	2088	2024	2017
Cl-	1765	1793	1845	1841	2098	2088	2023	2016
F-	1767	1790	1843	1841	2097	2087	2023	2016
H-	1767	1790	1844	1840	2097	2087	2023	2016
CH$_3$-	1762	1785	1841	1838	2097	2086	2022	2015
CH$_3$O-	1759	1781	1837	1834	2096	2086	2018	2012

There is a linear dependence between experimental (v_{exp}) and calculated by the PM6 method (v_{calc}) values of the normal vibrations frequencies of carbonyl groups of the benzoyl peroxide derivatives (Fig. 11.3).

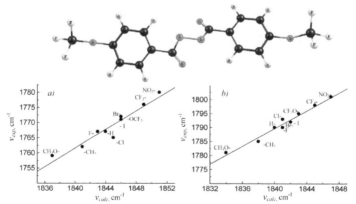

FIGURE 11.3   Structural model of (4-CF$_3$O-C$_6$H$_4$-COO)$_2$ (PM6 method) and dependence of the normal antisymmetric (a) and symmetric (b) vibration frequencies of carbonyl groups obtained experimentally (v_{exp}) and calculated by the PM6 method (v_{calc}) for the benzoyl peroxide derivatives.

These linear dependences (Fig. 11.3) in the linear regression analysis are described by the following equations:

$$V_{exp} = (-1072 \pm 297) + (1.54 \pm 0.16) \cdot V_{calc}, n = 10, R = 0.959 \qquad (4)$$

$$V_{exp} = (-1141 \pm 187) + (1.59 \pm 0.10) \cdot V_{calc}, n = 10, R = 0.984 \qquad (5)$$

Thus, among of the used semiempirical methods only PM6 provides a mathematical model for predicting of the normal vibrations frequencies position of the carbonyl group in the IR spectrum of the benzoyl peroxide symmetric derivatives.

The positions of the normal vibrations mode of carbonyl groups can be used as an analytical signal of qualitative and quantitative analysis of this class of peroxides by IR-spectroscopy. DFT methods are promising for computer structural chemistry of peroxide compounds because they with sufficient accuracy reproduce parameters of the molecular geometry and electronic structure of peroxides [5]. The opportunity of different DFT methods to reproduce the values of the normal vibrations frequencies of the benzoyl peroxide C = O group was estimated. The calculations were performed using the 6-311G (d, p) basis set in all cases. The results are listed in Table 11.3.

TABLE 11.3 The Effect of the DFT Level on the Normal Vibrations Frequencies of the Benzoyl Peroxide C = O Group

Method	V_{as}, cm^{-1}	V_s, cm^{-1}	ΔV^*, cm^{-1}	ΔV_{as}^{**}, cm^{-1}	ΔV_s^{***}, cm^{-1}
Experiment [4]	1767	1790	23	0	0
BLYP	1739	1762	23	28	28
B1LYP	1839	1864	25	72	74
B3LYP	1827	1852	25	60	62
O3LYP	1832	1853	21	65	63
X3LYP	1834	1859	25	67	69
B971	1837	1862	25	70	72
B972	1857	1888	31	90	98
PBEH1PBE	1868	1894	26	101	104
PBE1PBE	1868	1894	26	101	104
BH and HLYP	1926	1953	27	159	163
M06HF	1938	1966	28	171	176

$$*\Delta v = v_s - v_{as}$$

$$**\Delta V_{as} = V_{as}^{calc} - V_{as}^{exp}$$

$$***\Delta V_s = V_s^{calc} - V_s^{exp}$$

An appropriate correspondence of the calculated and experimental values of $*\Delta v = v_s - v_{as}$ is observed for all DFT methods under consideration. The best reproduction of the normal vibrations frequencies of C = O group is observed in the case of the BLYP method.

11.4   CONCLUSIONS

IR spectra of the benzoyl peroxide and its symmetrical derivatives (4-R-PhCOO)$_2$, R: NO$_2$–, –CF$_3$–, CF$_3$O–, I–, Br–, Cl–, F–, H–, CH$_3$–, CH$_3$O– have been calculated by the semiempirical and DFT methods. There is a linear dependence between the frequencies of normal vibrations experimentally obtained and calculated (PM6, PDDG and AM1) spectra of the benzoyl peroxide. Among the used semiempirical methods, only PM6 provides an appropriate mathematical model for predicting of the position of the carbonyl group normal vibrations frequencies in the IR spectra of the benzoyl peroxide symmetrical derivatives. The values of the normal vibrations frequencies of C = O group of the benzoyl peroxide were calculated on the different DFT levels. It was found that the best reproduction of the experimental values was observed in the case of BLYP calculation method with the 6-311G (d, p) basis set.

KEYWORDS

- **Benzoyl peroxide**
- **DFT-methods**
- **Diacyl peroxides**
- **Infrared spectra**
- **Molecular modeling**
- **Semiempirical methods**

REFERENCES

1. Guillén, M. D., & Cabo, N. (2002). Fourier transform infrared spectra data versus peroxide and anisidine values to determine oxidative stability of edible oils. *Food Chemistry*. 77, 503–510.

2. Luk'anets, V. M., Zhukovskij, V. Ya., Tsvetkov, N. S., & Ginzburg, I. M. (1973). Issledovanie struktury perefirov i diatsil'nych perekisej alifaticheskih karbonovyh kislot metodom IK-spectroskopii (Investigation of peresters and diacyl peroxides of aliphatic hydrocarbon acids structure by IR spectroscopy method). *Zhurnal Teoreticheskij I Eksperimental'noj Khimii.*, *9*, 131–134.

3. Bellamy, L. I., Connelly, B. R., Philpotts, A. R., & Williams, R. L. (1960). Infrared spectra of anhydrides and peroxides Z. fur Elektrochem. *64*, 563–566.

4. Z'at'kov, I. P., Sagaidachnyj, D. I., & Zubareva, M. M. (1984). Kolebatel'nye spektry diatsyl'nyh peroksidov i perefirov (Vibrational spectra of diacyl peroxides and peresters) Universitetskoe: Minsk.

5. Young, D. (2001). Computational Chemistry: A Practical Guide for Applying Techniques to Real World Problems Wiley Interscience: New York.

6. Head-Gordon, M., Pople, A. J., & Frisch, M. J. (1988). MP2 energy evaluation by direct methods Chemical Physics Letters. *153*, 503–506.

7. Catoire, V., Lesclaux, R., Schneider, W. F., & Wallington, T. J. (1996). Kinetics and Mechanisms of the Self-Reactions of CCl_3O_2 and $CHCl_2O_2$ Radicals and Their Reactions with HO_2. *J. Phys. Chem. 100*, 14356–14371.

8. Oxley, J., Smith, J., Brady, J., Dubnikova, F., Kosloff, R., Zeiri, L., & Zeir, Y. (2008). Raman and Infrared Fingerprint Spectroscopy of Peroxide-Based Explosives Society for Applied Spectroscopy. *62*, 906–915.

9. Gaussian, 09., Revision, A., Frisch, M. J., Trucks, G. W., Schleg el, H. B., Scuseria, G. E., Robb, M. A., Cheeseman, J. R., Scalmani, G. Barone, V. Mennucci, B. Petersson, G. A., Nakatsuji, H., Caricato, M. Li., X. Hratchian, H. P., Izmaylov, A. F., Bloino, J. Zheng, G. Sonnenberg, J. L., Hada, M., Ehara, M. Toyota, K. Fukuda, R. Hasegawa, J. Ishida, M. Nakajima, T. Honda, Y. Kitao, O. Nakai, H., Vreven, T., Jr. Montgomery, J. A., Peralta, J. E., Ogliaro, F., Bearpark, M., Heyd, J. J., Brothers, E., Kudin, K. N., Staroverov, V. N., Kobayashi, R., Normand, J., Raghavachari, K., Rendell, A., Burant, J. C., Iyengar, S. S., Tomasi, J., Cossi, M., Rega, N., Millam, J. M., Klene, M., Knox, J. E., Cross, J. B., Bakken, V., Adamo, C., Jaramillo, J., Gomperts, R., Stratmann, R. E., Yazyev, O., Austin, A. J., Cammi, R., Pomelli, C., Ochterski, J. W., Martin, R. L., Morokuma, K., Zakrzewski, V. G., Voth, G. A., Salvador, P., Dannenberg, J. J., Dapprich, S., Daniels, A. D., Farkas, O., Foresman, J. B., Ortiz, J. V., Cioslowski, J., & Fox, D. J. (2009). Gaussian, Inc., Wallingford, CT.

10. Dewar, M. J. S., Zoebisch, E. G., & Healy, E. F. (1985). AM1: A New General Purpose Quantum Mechanical Molecular Model. *J. Am. Chem. Soc., 107*, 3902–3909.

11. Stewart, J. J. P. (2007). Optimization of parameters for semiempirical methods. V. Modification of NDDO approximations and application to 70 elements. *J. Mol. Model., 13*, 1173–1213.

12. Tirado-Rives, J., & Jorgensen, W. L. (2008). Performance of B3LYP density functional methods for a large set of organic molecule. *J. Chem. Theory and Comput. 4*, 297–306.

13. Miehlich, B., Savin, A., Stoll, H., & Preuss, H. (1989). Results obtained with the correlation-energy density functionals of Becke and Lee, Yang and Parr. *J. Chem. Phys. Lett. 157*, 200–206.

14. Adamo, C., & Barone, V. (1997). Toward reliable adiabatic connection models free from adjustable parameters. *J. Chem. Phys. Lett. 274*, 242–250.

15. Hamprecht, F. A., Cohen, A., Tozer, D. J., & Handy, N. C. (1998). Development and assessment of new exchange-correlation functional. *J. Chem. Phys. 109*, 6264–6271.

16. Wilson, P. J., Bradley, T. J., & Tozer, D. J. (2001). Hybrid exchange-correlation functional determined from thermochemical data and ab initio potentials. *J. Chem. Phys. 115*, 9233–9242.

17. Becke, A. D. (1993). A new mixing of Hartree-Fock and local density-functional theories. *J. Chem. Phys. 98*, 1372–1377.

18. Zhao, Y., & Truhlar, D. G. (2006). Density Functional for Spectroscopy: No Long-Range Self-Interaction Error, Good Performance for Rydberg and Charge-Transfer States, and Better Performance on Average than B3LYP for Ground States. *J. Phys. Chem. A., 110*, 13126–13130.

19. Cohen, A. J., & Handy, N. C. (2001). *Dynamic correlation Mol. Phys., 99*, 607–615.

20. Adamo, C., & Barone, V. (1999). Toward reliable density functional methods without adjustable parameters: The PBE0 model. *J. Chem. Phys. 110*, 6158–6169.

21. Ernzerhof, M., & Perdew, J. P. (1998). Generalized gradient approximation to the angle- and system-averaged exchange hole. *J. Chem. Phys. 109*, 3313–3320.

22. Xu, X., & Goddard, W. A. (2004). The X3LYP extended density functional for accurate descriptions of nonbond interactions, spin states, and thermochemical properties. *Proc. Natl. Acad. Sci. USA., 101*, 2673–2677.

23. Barone, V., & Cossi, M. (1998). Quantum Calculation of Molecular Energies and Energy Gradients in Solution by a Conductor Solvent Model. *J. Phys. Chem. A., 102*, 1995–2001.

24. www.chemcraftprog.com

25. Sax, M., & McMullan, R. K. (1967). *The Crystal Structure of Dihenzoyl Peroxide and the Dihedral Angle in Covalent Peroxides Acta Cryst., 22*, 281–289.

26. Antonovskij, V. L., & Khursan, S. L. (2003). Fizicheskaia khimia organicheskih peroksidov (*Physical chemistry of organic peroxides*) PTC "AKADEMKNIGA": Moskva.

CHAPTER 12

POLYELECTROLYTE ENSYM-BEARING MICRODIAGNOSTICUM: A NEW STEP IN CLINICAL-BIOCHEMISTRY ANALYSIS

SERGEY A. TIKHONENKO, ALEXEI V. DUBROVSKY,
EKATERINA A. SABUROVA, and LYUDMILA I. SHABARCHINA

CONTENTS

12.1   INTRODUCTION

Clinical and biochemical analyzes are among the most common methods used to diagnose human diseases. Investigations of this kind are known to include general blood and urine tests, the study of a number of other biological fluids [1–3]. Until recently, these tests were carried out by chemical methods, but due to the toxicity of many of them, the low sensitivity and other shortcomings, enzymological methods widespread received today. However, along with the obvious advantages of this approach, there are some drawbacks: the ambiguity of the analysis in the presence of aggressive high-molecular compounds to the enzyme, in particular, proteases and other intracellular components; one-time use of the enzyme, etc. Thus, there is a need to protect the enzyme from the adverse effects, while maintaining access to it of the substrate, to increase its stability during prolonged storage, as well as to develop the reusability of the enzyme [4]. One type of such a defense is encapsulation of enzymes in polyelectrolyte microcapsules (PMC) [5, 6].

The PMC are the product of a new field of polymer nanotechnology. At present this area is booming around the world: in the U.S., EU, China, Australia, and another country. This, along with pure fundamental research of structure, physico-chemical and biological properties of polyelectrolyte microcapsules, increasing emphasis on applied research aimed at practical use of PMC, particularly in medicine, chemical engineering, biotechnology and many other areas [7]. The combination of unique properties and relatively simple technology of preparation for a wide range of PMC with the given parameters (structural, mechanical, functional), simplicity of inclusion of a wide variety of substances, and the ability to control membrane permeability PMC makes promising their use as tools for targeted drug delivery to the organs and tissues, [8–13], depot [8] and the therapeutic effect, as microreactors, microcontainers.

PMC containing enzyme can be used as a microdiagnosticum to detect a substrate, or inhibitor, or activator of the encapsulated enzyme in native biological fluids and in wasted water.

12.2   EXPERIMENTAL PART

12.2.1   REAGENTS

Pig skeletal muscle lactate dehydrogenase (EC 1.1.1.27) (M4 isoform) was isolated as in [14]. Urease (EC 3.5.1.5, jack bean) and proteinase K (EC 3.4.21.64) were from Fluka (no. 94–285) and Sigma (P8044), respectively. The initial reagents used in the preparation of the samples were as follows: sodium poly(styrenesulfonate) (70kDa), poly(diallyl dimethyl ammonium chloride) (70kDa), and poly (allylamine hydrochloride) (70kDa) from Aldrich (Germany); dextran sulfate (10kDa) and EDTA from Sigma (Germany); and calcium chloride, sodium carbonate, and sodium chloride from Reakhim (Russia). The polyelectrolyte solutions were brought to a specified pH value with concentrated solutions of NaOH or HCl.

12.2.2 CACO₃ MICROSPHERULITES CONTAINING ENZYMES

$CaCO_3$ Microspherulites Containing Enzymes were prepared as in Ref. [6]. One volume of a 1 M $CaCl_2$ aqueous solution was added to two volumes of enzyme solution (1.0 mg/mL) with vigorous stirring. Then an equal volume of aqueous 0.33 M Na_2CO_3 was rapidly added. After stirring for 30 s, the suspension of particles was allowed to stand for 15 min at room temperature without stirring. The formation of microspherulites was controlled with a light microscope. These composite microspherulites with inclusions of urease or LDH had a narrow size distribution with a mean diameter of ~4.5 μm.

12.2.3 PREPARATION OF PEMC

Involved adsorption of alternating layers of polyanions (PSS, DS) and polycations (PAA, PDADMA) on the surface of the $CaCO_3$-enzyme composite microspherulites in solutions containing 0.5 M NaCl and polyelectrolytes at 2 mg/mL. After each adsorption stage, the system was triply washed in order to remove unbound polymer molecules. After depositing a necessary number of polyelectrolyte layers, the carbonate core was dissolved in 0.2 M EDTA for 12h. The microcapsules thus prepared were triply washed with water; the enzyme remained inside the microcapsules. All procedures were performed at 20 °C.

12.2.4 UREASE ACTIVITY

It was determined from the decomposition of urea into ammonia and carbon dioxide CO_2 with a pH-sensitive dye Bromocresol Violet. The reaction mixture contained a necessary amount of urea and 0.015 mM Bromocresol Violet adjusted to pH 6.2. The known number of PEMC containing urease was added to the solution. Then, the kinetics of the reaction was measured from a change in absorption of the dye at 588 nm [15]. The activity of the free enzyme in the presence of the polyelectrolyte was measured by a similar method; a necessary amount of the polyelectrolyte with pH adjusted to 6.2 was added to the reaction solution.

12.2.5 LACTATE DEHYDROGENASE ACTIVITY

It was determined from a change in NADH absorption at 340 nm with a Specord M-40 spectrophotometer (Carl Zeiss, Germany) [15]. The reaction was initiated by introducing 20 μL of the enzyme (0.05–0.10 mg/mL) or a necessary amount of microcapsules with the enzyme into the reaction mixture containing pyruvate at a specified concentration, 0.2 mM NADH, and, in a number of cases, the polyelectrolyte.

12.2.6 STABILITY OF ENCAPSULATED ENZYMES AGAINST PROTEINASE K

We prepared two samples, each containing an aqueous suspension (0.5 ml) of microcapsules with urease (7.4×10^7 capsules/mL) in 0.01 M Tris-HCl (pH 8.0). A solution (0.5 mL) of proteinase K at 4 mg/mL in the same buffer was added to the first sample, and the same amount of the buffer was added to the second sample. Then, the solutions were incubated at 37 °C. Every 10 min, 50-µL portions were taken and added to the urease assay mixture (1.95 ml) with 125 mM urea. Free urease at 20 µg/mL instead of microcapsules was used as a reference.

12.3 RESULTS AND DISCUSSION

12.3.1 SHELL FORMATION WITH A SELECTED PAIR OF POLYELECTROLYTES

Since incorporation of enzymes into capsules of any origin usually alters the enzyme activity, it is most important to select a pair of oppositely charged polyelectrolytes that would be optimal for the operation of the projected microdiagnostic. Figure 12.1 shows the dependence of the enzyme activity on the concentration of different polyelectrolytes.

The PSS polyanion with a hydrophobic backbone is a strong inhibitor for LDH, whereas the PAA polycation with a highly hydrophobic backbone is a strong inhibitor for urease. A 50% inhibition of LDH and urease activity is observed at low concentrations of the polyelectrolytes (fractions of a microgram per milliliter). An important factor responsible for the inhibitory effect of the polyelectrolytes is the presence of the hydrophobic polymer backbone. Polyelectrolytes with a hydrophilic backbone (e.g., DS) have virtually no effect on the activity of the enzymes (Fig. 12.1a). However, the capsules prepared from these polyelectrolytes are not stable. The activity of LDH remains unchanged even in the presence of the studied polyelectrolytes with positively charged ionogenic groups up to very high concentrations. Therefore, polyelectrolytes that can be used for the preparation of polyelectrolyte microcapsules containing enzymes must satisfy the following criteria: first, these polyelectrolytes must not inactivate the enzyme at concentrations suitable for the formation of the capsule shell (1–2 mg/mL), and, second, they must have a hydrophobic backbone. It is these two criteria (sometimes incompatible) that determine the proper choice of a pair of oppositely charged polyelectrolytes for use in the design of polyelectrolyte enzymatic microdiagnostics.

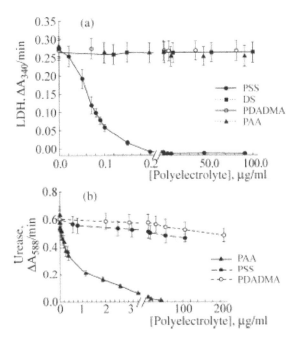

FIGURE 12.1 Dependence of (a) LDH and (b) urease activity on the concentration of polyelectrolytes (legend). LDH assay: LDH, 0.5 mg/mL; pyruvate, 1 mM; NADH, 0.2 mM; Tris-HCl, 0.05 M; pH 6.2. Urease assay: urease, 0.5 mg/mL; urea, 125 mM; Bromocresol Violet, 0.015 mM; pH 6.2.

The experiments and calculations have demonstrated that the best pairs of polyelectrolytes for the shells of microcapsules are PAA/DS and PAA/PSS for LDH, and PSS/PAA and PSS/PDADMA for urease in the given sequence of layer deposition. On this basis, we designed and prepared PEMC containing LDH and urease with different polyelectrolyte compositions and different numbers of layers.

12.3.2 'THE DEPENDENCE OF PROTEINS' DISTRIBUTION WITHIN POLYELECTROLYTE MICROCAPSULES ON PH OF THE MEDIUM

Transmission electron microscopy of ultrathin sections and confocal laser scanning microscopy were used to study the distribution of proteins within polyelectrolyte microcapsules. Since obtaining quality images using transmission electron microscopy is possible in case studies of electron dense objects, the iron containing protein, ferritin, with a value of pI 4.7, was selected for encapsulation. Polyelectrolytes polyallylamine (PAA) and polystyrene (PSS) that were used have values of ionization constants of functional groups of 10.5 and 1.9 respectively, which removes the question of their charge in the range of pH investigated. Photos of ultrathin sections of samples containing polyelectrolyte microcapsules ferritin at pH values below or near the isoelectric point are shown in Fig. 12.2. It was found that the distribution of proteins inside the capsule depends on the protein isoelectric point and the charge of the internal polyelectrolyte layer. The distribution of protein in the interior of the capsule is available in two versions: a uniform distribution of protein throughout the volume and concentration of its aggregates in wall space (Fig. 12.2).

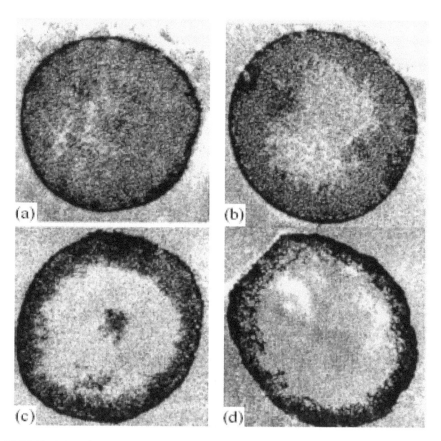

FIGURE 12.2   Electron micrographs of ultrathin sections of polyelectrolytes microcapsules containing ferrit in at pH 2 (a), 3 (b), 4 (c), 5 (d). The polyelectrolyte shell of the composition is (PAA/PSS) 3.

Thus, if the pH is less than the isoelectric point (pI 4.7), a protein is positively charged, while the inner layer of polyelectrolyte microcapsules is presented as a polycation, the protein molecules are distributed throughout its volume. Protein molecules lose their charge values near the isoelectric point and are concentrated in the wall space of the capsule due to hydrophobic interactions with a polyelectrolyte shell. If the polyanion PSS was used as the first layer in the formation of a shell, the protein at all pH values in the range studied was located in the wall space (Fig. 12.3). We attribute this to the electrostatic interaction between protein molecules and the polyelectrolyte at low pH and hydrophobic interactions in the region of the isoelectric point.

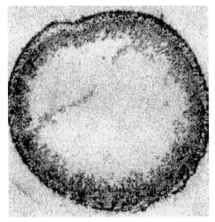

FIGURE 12.3   Electron micrographs of ultrathin sections of polyelectrolytes microcapsules containing ferrit in at pH 2. The polyelectrolyte shell of the composition is (PSS/PAA) 3.

12.3.3   ACTIVITY AND CATALYTIC CHARACTERISTICS OF ENCAPSULATED ENZYMES

According to the results presented in the earlier section, the microcapsules containing LDH were prepared so that the first layer of their shell was a positively charged polyelectrolyte, PAA, whereas the second and subsequent even layers were composed of different negatively charged polyelectrolytes (DS or PSS). In the microcapsules containing urease, the first layer of their shell was a negatively charged polyelectrolyte, PSS, whereas the second and subsequent even layers were composed of different positively charged polyelectrolytes (PAA or PDADMA). This was done to preventing or diminishes direct contact of the inactivating polyelectrolytes with the enzyme. It can be seen in Fig. 12.4a that LDH activity in the (PAA/PSS) 3 microcapsule is relatively low. Therefore, the inhibitory effect of PSS on the activity of LDH manifests itself even in the case where the PSS layer is formed after the PAA layer. Microcapsules composed of different bilayers, for example, (PAA/DS) 2(PAA/PSS), exhibit relatively high activity of the enzyme. However, the microcapsules in which the negatively charged polyelectrolyte in the shell is represented only by DS have zero activity. The reason for this is that DS has a polar sugar back-bone; moreover, although DS does not inhibit LDH, in the pair with PAA it forms a loose unstable shell. Eventually, a shell of the composition (PAA/DS) 2 (PAA/PSS), which involves polyelectrolytes with polar and hydrophobic backbones, was found to be optimal for encapsulation of LDH. Similarly, the inhibitory effect of the positively charged polyelectrolyte on the activity of urease was taken into account in the design and preparation of the microcapsule shell for the urease microdiagnostic. The urease-containing microcapsules (Fig. 12.4b) in which the first layer is a "non-inhibitory" polyelectrolyte, PSS, exhibit sufficiently high activity.

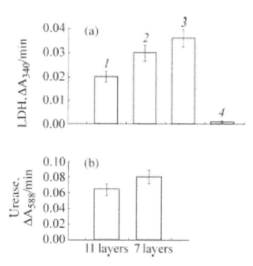

FIGURE 12.4   Activity of (a) LDH and (b) ureases in microcapsules with different compositions and different number of layers. LDH capsules: (*1*) (PAA/PSS) 3, (*2*) (PAA/DS) (PAA/PSS) 2, (*3*) (PAA/DS) 2(PAA/PSS), and (*4*) (PAA/DS) 3. Urease capsules: (PSS/PAA) 3 PSS (seven layers) and (PSS/PAA) 5 PSS (11 layers). Assay conditions as in Fig. 12.2, with 1.2′106 capsules/mL.

In order to determine the catalytic characteristics of the encapsulated enzymes, we obtained the dependences of the stationary rate of substrate conversion on the substrate concentration. As an example, the curves of saturation of urease with urea in the reaction of urea decomposition are depicted in Fig. 12.5. It can be seen that the dependences for urease in microcapsules, are generally similar to those for free en-zyme, except small differences in the affinity constants. In particular, the Michaelis constant K_M with respect to urea is 7.1×2.2 mM for urease in microcapsules of 11 and seven layers, whereas the K_M for free urease is 2.5×0.7 mM. The maximal rate Vmax for urease in microcapsules of 11 layers is 20% lower than that for urease in microcapsules of seven layers. The KM with respect to pyruvate for microcapsules containing LDH was not different from for free enzyme.

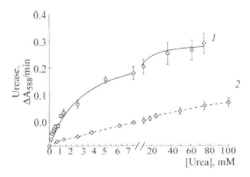

FIGURE 12.5   Dependences of the activity of (*1*) free urease and (*2*) urease in (PSS/PAA) 3 PSS microcapsules on the urea concentration.

12.3.4   ACTIVITY OF FREE AND ENCAPSULATED ENZYMES DURING LONG-TERM STORAGE AND UNDER THE ACTION OF PROTEINASE K

It is known that enzymes, especially at low concentrations, are inactivated during storage in solution. For oligomeric enzymes, this is associated with their dissociation into subunits and structural distortion of the active site. In the general case, inactivation of different enzymes in long-term storage can be caused by spontaneous thermal denaturation; chemical modification as a result of hydrolysis, oxidation, and other processes; and bacterial contamination. We investigated the stability of encapsulated LDH and urease, in the course of long-term storage. The dependences of the residual activity of the enzymes on the incubation time are plotted in Fig. 12.6. It can be seen that, unlike free enzymes in solution, the encapsulated enzymes retain their activity for several months. The investigation into the nature of the factors responsible for the high stability of an enzyme in a PEMC during long-term storage will be reported in a separate paper.

We also investigated the influence of proteolytic. Figure 12.6 shows the dependences of the activity of free (curve *1**) and encapsulated (curve *2**) urease on the time of incubation with proteinase K at 37 °C. For comparison, the dependences obtained in the absence of proteinase K (curves *1, 2*) are also shown in Fig. 12.6. Urease was encapsulated into the (PSS/PAA) 3 PSS shell. As can be seen in Fig. 12.7, the activity of free urease in the presence of the proteolytic enzyme steeply decreases to zero, whereas encapsulated urease retains the ability to decompose urea in the presence of proteinase. Since the enzymes in PEMC are not degraded by proteinase K, these microcapsules can be used for quantitative analysis of low-molecular compounds in biological fluids containing proteolytic enzymes. This obviates the need for removing proteinases from the biological fluid to be analyzed, which is a laborious process.

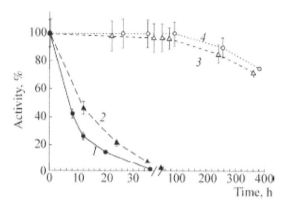

FIGURE 12.6   Variations in the activity of (*1*) free LDH, (*2*) free urease, (*3*) LDH in (PAA/DS) 2 (PAA/PSS) microcapsules, and (*4*) urease in (PSS/PAA) 3 PSS microcapsules during long-term storage in H_2O, $T = 21$ °C. LDH and urease, 50 mg/mL; microcapsules, $6.2'10^7$ mL^{-1}.

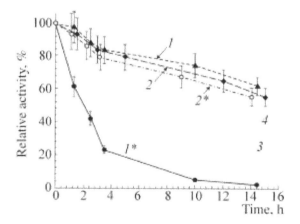

FIGURE 12.7   Dynamics of the activity of (*1*, *1**) free urease and (*2*, *2**) urease in (PSS/PAA) 3 PSS microcapsules at 37 °C: (*1**, *2**) in the presence and (*1*, *2*) in the absence of proteinase K.

12.4 CONCLUSIONS

Analysis of the data obtained for two enzymes belonging to different classes allows the following conclusions.

1. When the shell of the capsule is prepared from a properly chosen pair of polyelectrolytes, the encapsulated enzyme retains high affinity to the substrate.
2. The enzyme incorporated into a multilayered polyelectrolyte capsule retains activity for several months, whereas the activity of the free enzyme in solution drops nearly to zero in a few days.
3. The encapsulated enzyme completely retains activity in the presence of proteinase. This obviates the need for their removal from the biological fluid to be analyzed.
4. The proposed microdiagnostic based on encapsulated enzymes can be repeatedly used for analyzing biological fluids, thus further reducing the enzyme expenditure.
5. Such enzymic microdiagnostics represent a new class of biosensors with significant advantages over the existing clinical/biological enzymatic methods of analysis.
6. The distribution of proteins inside polyelectrolyte microcapsules depends on charges of protein and of inner polyelectrolyte layer.

REFERENCES

1. Dolgov, V. V., & Selivanova, A. V. (2006). Bio chemical studies in clinical diagnostic laboratories primary care health services. St. Petersburg "Vital Diagnostics SPb," 231.
2. Dolgov, V. V., Shevchenko, O. P., Sharyshev, A. A., & Bondar, V. A. (2007). Turbodimetriya in laboratory practice M. Reaform, 176.
3. Menshikova, V. V. (2009). Methods of clinical laboratory tests Moscow: Labora. 304.
4. Caruso, F., Trau, D., Mohwald, H., & Renneberg, R. (2000). Enzyme Encapsulation in Layer-by-Layer Engineered Polymer Multilayer Capsules. *Langmuir, 16,* 1485–1488.
5. Decher, G., Hong, J. D., & Schmitt, J. (1992). Buildup of ultrathin multilayer films by a self-assembly process 3. Alternating adsorption of anionic and cationic poly electrolytes on charged surfaces. *Thin Solid Films, 210(1–2),* 831–835.
6. Petrov, A. I., Volodkin, D. V., & Sukhorukov, G. B. (2005). Protein calcium carbonate coprecipitation: a tool of protein encapsulation. *Biotechnol Prog, 21(3),* 918–925.
7. Skirtach, A. G. et al. (2007). Nanoparticles distribution control by polymers: Aggregates versus nonaggregates. *Journal of Physical Chemistry C, 111(2),* 555–564.
8. Rivera-Gil, P. et al. (2009). Intracellular Processing of Proteins mediated by biodegradable polyelectrolyte capsules. *Nano Lett, 9(12),* 4398–4402.
9. Sabini, E. et al. (2008). Structural Basic for substrate promiscuity of dCK, *J. Mol Biol, 378(3),* 607–721.

10. Sieker, F. et al. (2008). Differential tapasin dependence of MHC class I molecules correlates with conformational changes upon peptide dissociation: a molecular dynamics simulation study. *Mol Immunol, 45(14),* 3714–3722.
11. Sieker, F., Springer, S., & Zacharias, M. (2007). Comparative molecular dynamics analysis of tapas in dependent and independent MHC class I alleles. *Protein Science, 16(2)* 299–308.
12. Borodina, T. N., Rumsh, L. D., Kunizhev, S. M., Sukhorukov, G. B., Vorozhtsov, G. N., Feldman, B. M., Rusanova, A. V., Vasilyeva, T. V., Strukova, C. M., & Markvicheva, E. A. (2007). The inclusion of extracts of medicinal plants in biodegradable microcapsules. *Biomédical Chemistry, 53(6),* 662–671.
13. Borodina, T. N., Rumsh, L. D., Kunizhev, S. M., Sukhorukov, G. B., Vorozhtsov, G. N., Feldman, B. M., & Markvicheva, E. A. (2007). Polyelectrolyte microcapsules as delivery systems of bioactive substances. *Biomédical Chemistry, 53(5),* 557–565.
14. Saburova, E. A., Bobreshova, M. E., Elfimova, L. I., & Sukhorukov Biokhimiya, B. I. (2000). (Moscow) *65(8),* 1151 [*Biochemistry* (Moscow) *65(8)*, 976].
15. Karpishchenko (2002). *Medical Laboratory Technologies: A Handbook,* (Intermedika, St. Petersburg, [in Russian].

CHAPTER 13

A RESEARCH NOTE ON TRANSPORT PROPERTIES OF FILMS OF CHITOSAN AMIKACIN

A. S. SHURSHINA, E. I. KULISH, and S. V. KOLESOV

CONTENTS

ABSTRACT

Transport properties of films on a basis chitosan and medicinal substance are investigated. Sorption and diffusive properties of films are studied. Diffusion coefficients are calculated. Kinetic curves of release of the amikacin, having abnormal character is shown. The analysis of the obtained data showed that a reason for rejection of regularities of process of transport of medicinal substance from chitosan films from the classical fikovsky mechanism are structural changes in a polymer matrix, including owing to its chemical modification at interaction with medicinal substance.

13.1 INTRODUCTION

Systems with controlled transport of medicines are the extremely demanded [1, 2]. Research of regularities of processes of diffusion of water and medicinal substance in polymer films and opportunities of control of release of medicines became the purpose of this chapter. As a matrix for the immobilization of drugs used naturally occurring polysaccharide chitosan, which has a number of valuable properties: non-toxicity, biocompatibility, high physiological activity [3], as well as a drug used aminoglycoside antibiotic series – amikacin, actively applied in the treatment of pyogenic infections of the skin and soft tissue [4].

13.2 EXPERIMENTAL PART

The object of investigation was a chitosan (ChT) specimen produced by the company "Bio progress" (Russia) and obtained by acetic deacetylation of crab chitin (degree of deacetylation ~84%) with $M_{sd}=334000$. As the medicinal substance (MS) used an antibiotic amikacin (AM) quadribasic aminoglycoside, used in the form of salts sulfate (AMS) and chloride (AMCh). Chemical formulas of objects of research and their symbols used in the text are given in Table 13.1.

TABLE 13.1   Formulas Research Objects and Symbols Used in the Reaction Schemes

Formula object of study	Symbol
Chitosan acetate monomer unit CH_2OH (structure) OH $NH_3^+CH_3COO^-$	$\sim\!\!\sim\!\!\sim NH_3^+CH_3COO^-$

Amikacin sulfate	
Amikacin chloride	

ChT films were obtained by means of casting of the polymer solution in 1% acetic acid onto the glass surface with the formation of chitosan acetate (ChTA). Aqueous antibiotic solution was added to the ChT solution immediately before films formation. The content of the medicinal preparation in the films was 0.01, 0.05 and 0.1 mol/mol ChT. The film thickness in all experiments was maintained constant and equal 100 microns. To study the release kinetics of MS the sample was placed in a cell with distilled water. Stand out in the aqueous phase AM recorded spectrophotometrically at a wavelength of 267 nm, corresponding to the maximum absorption in the UV spectrum of MS. Quantity of AM released from the film at time t (G_s) was estimated from the calibration curve. The establishment of a constant concentration in the solution of MS G_∞ is the time to equilibrium. MS mass fraction α, available for diffusion, assessed as the quantity of films released from the antibiotic to its total amount entered in the film.

Studying of interaction MS with ChT was carried out by the methods of IR- and UV-spectroscopy. IR-spectrums of samples wrote down on spectrometer "Shimadzu" (the tablets KBr, films) in the field of 700–3600 cm^{-1}. UV-spectrums of all samples removed in quartz ditches thick of 1 cm concerning water on spectrophotometer "Specord M-40" in the field of 220–350 nanometers.

With the aim of determining the amount of medicinal preparation held by the polymer matrix β there was carried out the synthesis of adducts of the ChT-antibiotic interaction in acetic acid solution. The synthesized adducts were isolated by double reprecipitation of the reaction solution in NaOH solution with the following washing of precipitated complex residue with isopropyl alcohol. Then the residue was dried in vacuum up to constant mass. The amount of preparation strongly held by chitosan matrix was determined according to the data of the element analysis on the analyzer EUKOEA–3000 and UF-spectrophotometrically.

The relative amount of water m_t absorbed by a film sample of ChT, determined by an execrator method, maintaining film samples in vapors of water before saturation, and calculated on a formula: $m_t = (\Delta m)/m_0$, where m_0 is the initial mass of ChT in a film, Δm is weight the absorbed film of water by the time of t time.

Isothermal annealing of film samples carried out at a temperature 120°C during fixed time. Structure of a surface of films estimated by method of laser scanning microscopy on device LSM-5-Exciter (Carl Zeiss, Germany).

13.3   RESULTS

It is well known that release of MS from polymer systems proceeds as diffusive process [5–7]. However a necessary condition of diffusive transport of MS from a polymer matrix is its swelling in water, that is, effective diffusion of water in a polymer matrix. Diffusing in a polymer matrix, water molecules, possessing it is considerable bigger mobility in comparison with high-molecular substance, penetrate in a polymer material, separating apart chains and increasing the free volume of a sample. The main mechanisms in water transport in polymer films are simple diffusion and the relaxation phenomena in swelling polymer. If transfer is caused mainly mentioned processes, the kinetics of swelling of a film is described by the equation [8].

$$m_t/m_\infty = kt^n, \tag{1}$$

where m_∞ is the relative amount of water in equilibrium swelling film sample, k is a constant connected with parameters of interaction polymer-diffuse substance, n is an indicator characterizing the mechanism of transfer of substance. If transport of substance is carried out on the diffusive mechanism, the indicator of n has to be close to 0.5. If transfer of substance is limited by the relaxation phenomena's – n > 0.5.

The parameter n determined for a film of pure ChT is equal 0.63 (i.e., >0.5) that is characteristic for the polymers, being lower than vitrification temperature [9]. This fact is connected with slowness of relaxation processes in glassy polymers. Values of equilibrium sorption of water and indicator n defined for film samples, passed isothermal annealing (a relaxation of nonequilibrium conformations of chains with reduction of free volume), and are presented in Table 13.2.

TABLE 13.2   Parameters of Swelling of Chitosan Films in Water Vapor

Composition of the film	The concentration of MS in the film, mol/mol ChT	Annealing time, min	$D_s^a \times 10^{11}$, cm²/sec	$D_s^6 \times 10^{11}$, cm²/sec	n	Q_∞, g/g ChT
ChT		15	37.2	37.0	0.50	2.50
		30	36.3	36.0	0.48	2.48
		60	15.3	14.1	0.44	2.47
		120	12.2	13.0	0.43	2.46
ChT-AMCh	0.01	30	8.5	4.7	0.42	1.84
		60	7.0	4.5	0.39	1.56
		120	5.8	4.2	0.37	1.42
	0.05	30	5.3	3.2	0.34	1.85
	0.1	30	4.4	3.1	0.32	1.48
ChT-AMS	0.01	30	6.9	2.9	0.34	1.58
		60	4.0	2.8	0.27	1.46
		120	2.6	2.3	0.25	1.31
	0.05	30	6.1	2.2	0.30	1.66
	0.1	30	2.7	1.9	0.27	1.07

Apparently from Table 13.2 data, carrying out isothermal annealing leads to that values of an indicator n decrease. Thus, if annealing was carried out during small time (15–30 min), the value n determined for pure ChT is close to 0.5. It indicates that transfer of water is limited by diffusion, and it is evidence that ChT in heat films is in conformational relaxed condition. In process of increase time of heating till 60–120 min, values of an indicator n continue to decrease that, most likely, reflects process of further restructuring of the polymer matrix, occurring in the course of film heating. That processes of isothermal annealing of ChT at temperatures ≥100 °C are accompanied by course of a number of chemical transformations, was repeatedly noted in literature [10, 11]. In particular, it is revealed that besides acylation reaction, there is the partial destruction of polymer increasing the maintenance of terminal aldehyde groups which reacting with amino groups; sew ChT macromolecules at the expense of formation of azomethine connections. In Ref. [12], the fact of cross-linking in the HTZ during isothermal annealing was confirmed by the study of the spin-lattice relaxation. In the values of equilibrium sorption isothermal annealing, however, actually no clue, probably owing to the low density of cross-links.

A similar result reduction indicator n is achieved when incorporated into a polymer matrix MS. As the data in Table 13.2, the larger MS entered into the film, the slower and less absorb water ChT. During isothermal annealing medicinal films effect enhanced. Such deviations from the laws of simple diffusion (Fick's law) and others researches have observed, explaining their strong interaction polymer with MS [13].

In the aqueous environment from ChT film with antibiotic towards to the water flow moving to volume of the chitosan, from a polymeric film LV stream is directed to water.

In Fig. 13.1, typical experimental curves of an exit of AM from chitosan films with different contents of MS are presented. All the kinetic curves are located on obviously expressed limit corresponding to an equilibrium exit of MS (G).

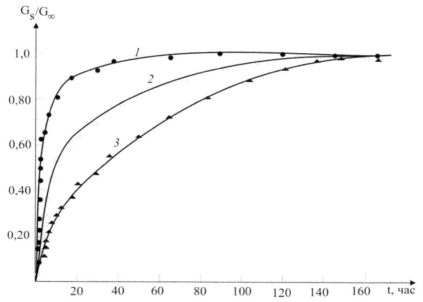

FIGURE 13.1 Kinetic curves of the releases of the MS from films systems ChT-AMCh with the molar ratio of 1:0.01 (1), 1:0.05 (2) and 1:0.1 (3). Isothermal annealing time 30 min.

Mathematical description of the desorption of low molecular weight component from the plate (film) with a constant diffusion coefficient is in detail considered in [14], where the original differential equation adopted formulation of the second Fick's law:

$$\frac{\partial c_s}{\partial_t} = D_s \frac{\partial^2 c_s}{\partial x^2}$$

(1)

The solution of this equation disintegrates to two cases: for big ($G_s/G_\infty > 0.5$) and small ($G_s/G_\infty \leq 0.5$) experiment times:

On condition $G_s/G_\infty \leq 0.5$ $\quad \dfrac{G_s}{G_\infty} = \left[16 D_s t / \pi L^2 \right]^{0.5}$

(2a)

On condition $G_s/G_\infty > 0.5$ $\quad \dfrac{G_s}{G_\infty} = 1 - \left[8 / \pi^2 \exp\left(-\pi^2 D_s t / L^2 \right) \right]$

(2b)

where $G_s(t)$ concentration of the desorbed substance at time t and G_∞-Gs value at $t \to \infty$, L–thickness of the film sample.

In case of transfer MS with constant diffusion coefficient the values calculated as on initial (condition $G_s/G_\infty \leq 0.5$), and at the final stage of diffusion (condition $G_s/G_\infty > 0.5$), must be equal. The equality $D_s^a = D_s^b$ indicates the absence of any complications in the diffusion system polymer-low-molecular substance [15]. However, apparently from the data presented in Table 13.3, for all analyzed cases, value of diffusion coefficients calculated at an initial and final stage of diffusion don't coincide.

TABLE 13.3 Desorption Parameters AMS and AMCh in Films Based on ChT

Composition of the film	The concentration of MS in the film, mol/mol ChT	Annealing time, min	$D_s^a \times 10^{11}$, cm^2/sec	$D_s^6 \times 10^{11}$, cm^2/sec	n	α
ChT-AMCh	1:0.01	30	81.1	3.3	0.36	0.95
		60	76.9	3.0	0.33	0.92
		120	37.0	2.9	0.25	0.88
	1:0.05	30	27.2	2.5	0.29	0.91
	1:0.1	30	25.9	2.2	0.21	0.83

TABLE 13.3   *(Continued)*

Composition of the film	The concentration of MS in the film, mol/mol ChT	Annealing time, min	$D_s^a \times 10^{11}$, cm²/sec	$D_s^6 \times 10^{11}$, cm²/sec	n	α
ChT-AMS	1:0.01	30	24.6	1.9	0.30	0.94
		60	23.0	1.8	0.22	0.90
		120	21.1	1.4	0.20	0.85
	1:0.05	30	24.0	1.8	0.18	0.80
	1:0.1	30	23.3	1.5	0.16	0.70

Note that in the process of water sorption similar regularities (Table 13.2) were observed. This indicates to a deviation of the diffusion of the classical type and to suggest the so-called pseudonormal mechanism of diffusion of MS from a chitosan matrix.

About pseudo normal type diffusion MS also shows kinetic curves, constructed in coordinates $G_s/G_\infty - t^{1/2}$ (Fig. 13.2). In the case of simple diffusion, the dependence of the release of the MS from film samples in coordinates $G_s/G_\infty - t^{1/2}$ would have to be straightened at all times of experiment. However, as can be seen from (Fig. 13.2) a linear plot is observed only in the region $G_s/G_\infty < 0.5$, after which the rate of release of the antibiotic significantly decreases.

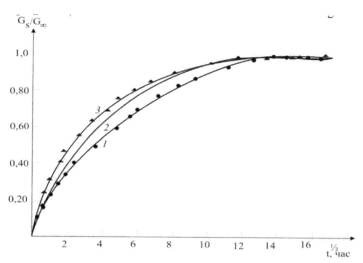

FIGURE 13.2   Kinetic curves of the release of the MS from film systems HTZ-AMS with a molar concentration of 0.01 (1) 0.05 (2) and 0.1 (3). Isothermal annealing time 30 min.

At diffusion MS from films, also as well as in case of sorption by films ChT–AM of water vapor, have anomalously low values of the parameter n, estimated in this case from the slope in the coordinates $\lg(G_s/G_\infty)$–lgt. Increase the concentration of MS and time of isothermal annealing, as well as in the process of sorption of water vapor films, accompanied by an additional decrease in the parameter n. symbiotically index n also changes magnitude α. Moreover, the relationship between the parameters of water sorption films ChT-AM (values of diffusion coefficient, an indicator n and values of equilibrium sorption Q) and the corresponding parameters of the diffusion of MS from the polymer film on condition $G_s/G_\infty \leq 0.5$ is shown.

All known types of anomalies of diffusion, can be described within relaxation model [16]. Unlike the fikovsky diffusion assuming instant establishment and changes in the surface concentration of sorbate, the relaxation model assumes change in the concentration in the surface layer on the first-order equation [17]. One of the main reasons causing change of boundary conditions is called nonequilibrium of the structural-morphological organization of a polymer matrix [16]. A possible cause of this could be the interaction between the polymer and LV.

Abnormally low values n in Ref. [18] were explained with presence tightly linked structure of amorphous-crystalline matrix. In this case, is believed, the effective diffusion coefficient in process of penetration into the volume of a sample can be reduced due to steric constraints that force diffusant bypass crystalline regions and diffuse to the amorphous mass of high-density cross-linking. As ChT belongs to the amorphous-crystalline polymers, it would be possible to explain low values n observed in our case similarly. However, according to Table 13.2 in the case of films of individual ChT such anomalies are not observed. Thus, the effect of substantially reducing n is associated with the interaction between HTZ and LV.

In the volume of the polymer matrix MS can be in different states. Part of the drug may be linked to the macromolecular chain through any chemical bonds on the other hand, some of it may be in the free volume in the form of physical filler. In the latter case, it can cause a certain structural organization of the polymer matrix. That antibiotic AM influences on the structure and morphology of the films ChT, indicate data of the laser scanning microscopy. As seen from the electron microscope images of Fig. 13.3, the initial films ChT visible surface strong interference caused by its heterogeneity. With the introduction of the film AM, the interference surface is significantly reduced.

IR-and UV-spectroscopy data indicate to taking place interaction between AM and ChT. It may be noted, for example, a significant change in the ratio between the intensity of the bands ChT corresponding hydroxyl and nitrogen-containing groups, before and after the interaction with the antibiotic (Table 13.4).

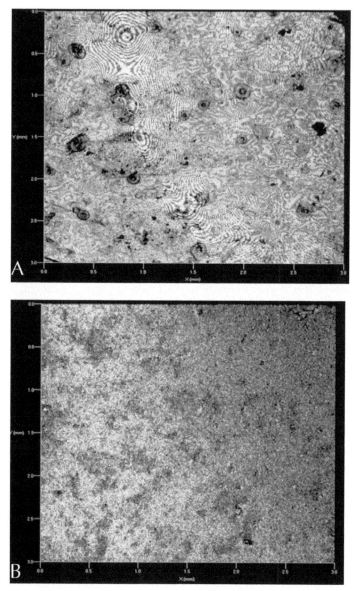

FIGURE 13.3   Micrograph of the surface (in contact with air) film individual ChT (a) and film ChT-AMS (b).

TABLE 13.4 The Value of the Intensity Ratio of the Absorption Bands of Some of the Data of the IR Spectra

Sample	I_{1640}/I_{2900}	I_{1590}/I_{2900}	I_{1458}/I_{2900}
ChT	0.57	0.51	0.72
AM	0.61	0.62	0.7
ChT-AM complex derived from 1% acetic acid	0.48	0.53	0.59

Binding energy in the adduct reaction ChT-antibiotic, estimated by the shift in the UV spectra of the order of 10 kJ/mol, which allows to tell about connection ChT-antibiotic by hydrogen bonds.

Thus, AM may interact with ChT by forming hydrogen bonds. However, interpretation of the data on the diffusion is much more important that AM can form chains linking ChT by salt formation.

Exchange interactions between ChT and AMS may occur under the scheme:

Due dibasic sulfuric acid, it is possible to suggest the formation of two types of salts, providing stapling ChT macromolecules with the loss of its solubility. Firstly, the water-insoluble "double" salt-sulfate ChT-AMS, secondly, the salt mixture insoluble in water ChT sulfates and soluble AM acetate.

If to take the AM in the form of chloride, an exchange reaction between ChT acetate and AM chloride reduces the formation of dissociated soluble salts. Accordingly, the reaction product in this case will consist of the H-complex ChT-AM.

Data on a share of antibiotic related to polymer adducts (β), obtained in solutions of acetic acid, and are presented in Table 13.5.

TABLE 13.5 Mass Fraction of the Antibiotic β, Defined in Reaction Adducts Obtained from 1% Acetic Acid

Used antibiotic	The concentration of MS in the film, mol/mol ChT	β
AMS	1.00	0.72
	0.10	0.33
	0.05	0.21
	0.01	0.07
AMCh	1.00	0.37
	0.10	0.20
	0.05	0.08
	0.01	0.04

As seen in Table 13.5, from the fact that AMC is able to "sew" chitosan chain is significantly more closely associated with macromolecules MS than for AMX.

Formation of chemical compounds of MS with ChT is probably the reason for the observed anomalies – reducing the rate of release of MS from film caused by simple diffusion, as well as the reduction of the share allocated to the drug (α). Indeed, the proportion of MS found in adducts of reaction correlates with the share of the antibiotic is not capable of participating in the diffusion process, and with the index n, reflecting the diffusion mechanism (Fig. 13.4).

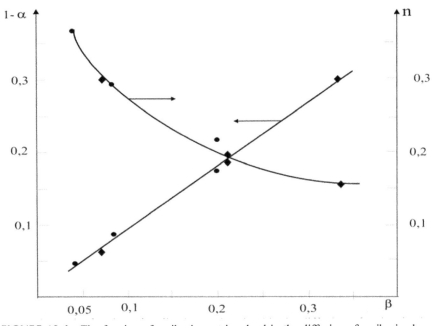

FIGURE 13.4 The fraction of amikacin, not involved in the diffusion of amikacin share determined in the adduct reaction. " -System ChT-AMS; ·-ChT-AMCh.

Thus, structural changes in the polymer matrix, including as a result of its chemical modification of the interaction with the drug substance, cause deviations regularities of transport MS of chitosan films from classic fikovskogo mechanism. Mild chemical modification, for example by cross-linking macromolecules salt formation, not affecting the chemical structure of the drug, is a possible area of control of the transport properties of medicinal chitosan films.

KEYWORDS

- **Chitosan**
- **Diffusion**
- **Medicinal substance**
- **Sorption**

REFERENCES

1. Shtilman, M. I. (2006). *Polymers of Medico Biological Appointment*, M: Akademkniga, 58.
2. Plate, N. A., & Vasilev, A. E. (1986). *Physiologically Active Polymers* M: Chemistry, 152.
3. Skryabin, K. G., Vihoreva, G. A., & Varlamov, V. P. (2002). Chitin and Chitosan, Preparation Properties and Application. M: Science, 365.
4. Mashkovskii, M. D. (1997). Pharmaceuticals Kharkov: Torsing *2*, 278.
5. Ainaoui, A., & Verganaud, J. M. (2000). *Comput. Theor. Polym. Sci., 10(2)*, 383.
6. Kwon, J. H., Wuethrich, T., Mayer, P., & Escher, B. I. (2009). *Chemosphere, 76*, 83.
7. Martinelli A., D'Ilario L., Francolini I., Piozzi A. (2011). *Int. J. Pharm. 407(1–2)*, 197.
8. Hall, P. J., Thomas, K., & Marsh, M. (1992). *Fuel 71(11)*. 1271.
9. Chalyih, A. E. (1987). *Diffusion in Polymer Systems* M: Chemistry, 136.
10. Zotkin, M. A., Vihoreva, G. A., Ageev, E. P. et al. (2004). *Engineering chemistry, 9*, 15.
11. Ageev, E. P., Vihoreva, G. A., Zotkin, M. A. et al. (2004). High-molecular compounds *46(12)*, 2035.
12. Smotrina, T. V. (2012). *Butlerovsky Messages 29(2)*, 98–101.
13. Singh, B., & Chauhan, N. (2008). *Acta Biomaterialia, 4(1)*, 1244.
14. Crank, J. (1975). *The Mathematics of Diffusion Oxford*: Clarendon Press 46.
15. Ukhatskaya, E. V., Kurkov, S. V., Matthews, S. E. et al. (2010). *Int. J. Pharm. 402(1–2)*, 10.
16. Malkin, A. Ya., & Chalyih, A. E. (1979). Diffusion and Viscosity of Polymers. Methods of Measurement. M: Chemistry, 304.
17. Pomerancev, A. L. (2003). Methods of nonlinear regression analysis to model the kinetics of chemical and physical processes Dissertation of the Doctor of Physical and Mathematical Sciences M: MGU.
18. Kuznecov, P. N., Kuznecova, L. I., & Kolesnikova, S. M. (2010). Chemistry in Interests of a Sustainable Development, *18*, 283–298.

CHAPTER 14

CRYSTALLINE PHASE MORPHOLOGY IN POLYMER/ORGANOCLAY NANOCOMPOSITES

K. S. DIBIROVA, G. V. KOZLOV, G. M. MAGOMEDOV, and G. E. ZAIKOV

CONTENTS

ABSTRACT

It has been shown that crystalline phase morphology in nano composites polymer/ organoclay with semicrystalline matrix defines the dimension of fractal space, in which the indicated nanocomposites structure is formed. In its turn, this dimension influences strongly on both deformational behavior and mechanical characteristics of nanocomposites.

14.1   INTRODUCTION

It has been shown earlier [1, 2], that particles (aggregates of particles) of filler (nanofiller) form network in polymer matrix, possessing fractal (in the general case multi fractal) properties and characterized by fractal (Hausdorff) dimension D_n. Hence, polymer matrix structure formation in nanocomposites can be described not in Euclidean space, but in fractal one. This circumstance tells to a considerable degree on both structure and properties of nano composites. As it has been shown in Ref. [2], polymer nano composites properties change is defined by polymer matrix structure change, which is due to nano filler introduction. So, the authors [3] demonstrated that the introduction of organoclay in high-density polyethylene (HDPE) resulted in matrix polymer crystalline morphology change, that is, in spherolites size increasing about twice. Therefore, this chapter's purpose is the study of polymer matrix crystalline morphology influence on structure and properties of nanocomposites high-density polyethylene/Na$^+$-montmorillonite (HDPE/MMT). This study was performed within the framework of fractal analysis [4].

14.2   EXPERIMENTAL PART

As a matrix polymer HDPE with melt flow index of approx. 1.0 g/10 min and crystallinity degree K of 0.72, determined by samples density, manufactured by firm Huntsman LLC, was used. As nanofiller organoclay Na$^+$-montmorillonite of industrial production of mark Cloisite 15, supplied by firm Southern Clay (USA), was used. A maleine anhydride (MA) was applied as a coupling agent. Conventional signs and composition of nano composites HDPE/MMT are listed in Table 14.1 [3].

Compositions HDPE/MA/MMT was prepared by components mixing on twin-screw extruder of mark Haake TW 100 at temperatures 390–410K. Samples with thickness of 1 mm were obtained on one-screw extruder Killion [3].

Uniaxial tension mechanical tests have been performed on apparatus Rheometric Scientific Instrument (RSA III) according to ASTM D882–02 at temperature 293 and strain rate 2 10^{-3} s^{-1}. The morphology of nanocomposites HDPE/MMT was studied with the help of polarized optical microscope Zeiss with magnification 40* and thus obtained spherolites average diameter D_{sp} in this way is also adduced in Table 14.1.

TABLE 14.1   Composition and Spherolites Average Diameter of Nanocomposites HDPE/MMT

Sample conventional sign	MA contents, mass%	MMT contents, mass%	Spherolites average diameter D_{sp}, mcm
A	–	–	5.70
B	1.0	–	5.80
C	–	1.0	11.62
D	–	2.5	10.68
E	–	5.0	10.75
F	1.0	1.0	11.68
G	2.5	2.5	11.12
H	5.0	5.0	11.0
I	5.0	2.5	11.30

14.3   RESULTS AND DISCUSSION

As it is known [5], the fractal dimension of an object is a function of space dimension, in which it is formed. In the computer model experiment this situation is considered as fractals behavior on fractal (but not Euclidean) lattices [6]. The space (or fractal lattice) dimension D_n can be determined with the aid of the following equation [1]:

$$v_F = \frac{2.5}{2 + D_n}, \tag{1}$$

where v_F is Flory exponent, connected with macromolecular coil dimension D_f by the following relationship [1]:

$$v_F = \frac{1}{D_f}, \tag{2}$$

In its turn, the dimension D_f value for linear polymers the following simple equation gives [7]:

$$D_f = \frac{2d_f}{3}, \tag{3}$$

where d_f is the fractal dimension of polymeric material structure, determined as follows [8]:

$$d_f = (d-1)(1+v),$$ (4)

where d is the dimension of Euclidean space, in which a fractal is considered (it is obvious, that in our case $d=3$), v is Poisson's ratio, estimated according to the results of mechanical tests with the help of the relationship [9]:

$$\frac{\sigma_Y}{E} = \frac{1-2v}{6(1+v)},$$ (5)

where σ_Y is yield stress, E is elasticity modulus.

In Fig. 14.1, the dependence of fractal space (fractal lattice) dimension D_n, in which the studied nanocomposites structure is formed, on spherolites mean diameter D_{sp} is adduced. As one can see, the correlation $D_n(D_{sp})$ is linear and described analytically by the following empirical equation:

$$D_n = 2.1 \times 10^{-2} D_{sp} + 2.5,$$ (6)

where value D_{sp} is given in mcm.

From the Eq. (6) it follows, that the minimum value D_n is achieved at $D_{sp}=0$ and is equal to 2.5, that according to the Eqs. (1)–(3) corresponds to structure fractal dimension $d_f \approx 2.17$. Since the greatest value of any fractal dimension for real objects, including D_n at that, cannot exceed 2.95 [8], then from the Eq. (6) the limiting value D_{sp} for the indicated matrix polymer can be evaluated, which is equal to ~ 21.5 mcm. Let us also note that the large scatter of the data in Fig. 14.1 is due to the difficulty of the value D_{sp} precise determination.

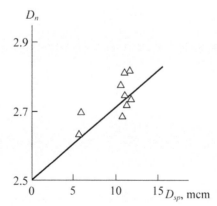

FIGURE 14.1   The dependence of space dimension D_n, in which nanocomposite structure is formed, on spherolites average diameter D_{sp} for nano composites HDPE/MMT.

As it is known [7] in semi crystalline polymers deformation process partial melting-recrystallization (mechanical disordering) of a crystalline phase can be realized, which is described quantitatively within the framework of a plasticity fractal theory [10]. According to the indicated theory Poisson's ratio value v_Y at a yield point can be evaluated as follows:

$$v_Y = \chi v + 0.5(1-\chi), \tag{7}$$

where χ is a relative fraction of elastically deformed polymer, v is Poisson's ration in elastic strains region, determined according to the Eq. (5), and v_Y value is accepted equal to 0.45 [7].

The calculation of a relative fraction of crystalline phase χ_{cr}, subjecting to mechanical disordering, can be performed according to the equation [7]:

$$\chi_{cr} = \chi - \alpha_{am} - \phi_{cl}, \tag{8}$$

where α_{am} is an amorphous phase relative fraction, which is equal to $(1-K)$, φ_{cl} is a relative fraction of local order domains (nano clusters), which can be determined with the aid of the following fractal relationship [7]:

$$d_f = 3 - 6 \times 10^{-10} \left(\frac{\phi_{cl}}{SC_\infty} \right)^{1/2}, \tag{9}$$

where S is cross-sectional area of macromolecule, which is equal to 14.4 Å^2 for HDPE [7], C_∞ is characteristic ratio, which is an indicator of polymer chain statistical flexibility [11], and connected with the dimension d_f by the following relationship [7]:

$$C_\infty = \frac{2d_f}{d(d-1)(d-d_f)} + \frac{4}{3}. \tag{10}$$

As it is known [7], parameter χ_{cr} effects essentially on deformational behavior and mechanical properties of semicrystalline polymers. In Fig. 14.2, the dependence $\chi_{cr}(D_n)$ is adduced for the studied nanocomposites, which turns out to be linear, that allows to describe it analytically as follows:

$$\chi_{cr} = 1.88(D_n - 2.55) \tag{11}$$

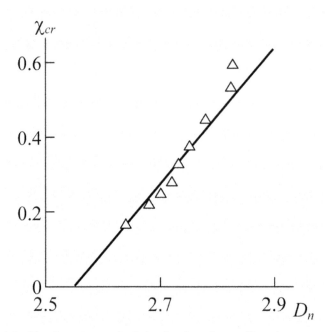

FIGURE 14.2   The dependence of relative fraction of crystalline phase χ_{cr}, subjecting to mechanical disordering, on space dimension D_n for nanocomposites HDPE/MMT.

From the Eq. (8) it follows that the greatest value χ_{cr} (χ_{cr}^{max}) is achieved at the following conditions: $\chi = 1.0$ and $\varphi_{cl} = 0$. In this case the condition $\chi_{cr}^{max} = K$ is realized, that was to be expected from the most common considerations. For the studied nanocomposites the value $\chi_{cr}^{max}=K=0.72$ is achieved according to the equation (11) at $D_n \approx 2.933$, that is close to the indicated above limiting value $D_n=2.95$ [8]. The minimum value $\chi_{cr}=0$ according to the Eq. (11) is achieved at $D_n=2.55$ or, according to the formulas (1)–(3), at $d_f=2.73$. As it is known [7], the value d_f can be determined alternatively as follows:

$$d_f \approx 2 + K \cdot \qquad (12)$$

From the Eqs. (11) and (12) it follows, that a common variation $\chi_{cr}=0$–0.72 is realized at the constant value $K=0.72$, that is, this parameter does not depend on crystallinity degree and it is defined only by polymer matrix crystalline morphology change.

As it is known [4], the parameter χ_{cr} influences essentially on nano composites polymer/organoclay properties. One from the most important mechanical characteristics of polymeric materials, namely, elasticity modulus E depends on the value χ_{cr} as follows:

$$E = \left(40 + 54.9\chi_{cr}\right)\sigma_Y. \tag{13}$$

In Fig. 14.3, the comparison of experimental E and calculated according to the Eq. (13) E^T elasticity modulus values for the considered nano composites is adduced. In this case the value χ_{cr} was determined according to the relationship (11). As it follows from the data of (Fig. 14.3) a theory and experiment good correspondence is obtained (the average discrepancy between E^T and E makes up ~ 14%), that will be enough for preliminary estimations performance.

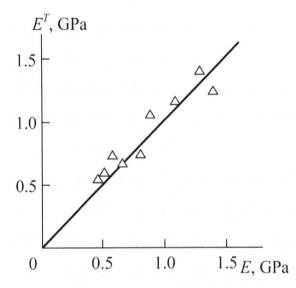

FIGURE 14.3 The relation between experimental E and calculated according to the Eqs. (11) and (13) E^T elasticity modulus values for nano composites HDPE/MMT.

14.4 CONCLUSIONS

Thus, this chapter results have shown that the fractal dimension of space, in which nano composites structure is formed, is defined by their crystalline polymer phase morphology and does not depend on crystallinity degree. The indicated dimension defines unequivocally partial melting-recrystallization process at nano composites deformation and influences strongly on their mechanical characteristics.

KEYWORDS

- **Elasticity modulus**
- **Fractal space**
- **Morphology**
- **Nanocomposite**
- **Organoclay**
- **Semicrystalline polymer**

REFERENCES

1. Kozlov, G. V., Yanovskii, Yu G., & Zaikov, G. E. (2010). *Structure and Properties of Particulate-Filled Polymer Composites: the Fractal Analysis* Nova Science Publishers, Inc., New York, 282.
2. Miktaev, A. K., Kozlov, G. V., & Zaikov, G. E. (2008). *Polymer Nanocomposites: Variety of Structural Forms and Applications* Nova Science Publishers, Inc., New York, 319.
3. Ranade, A., Nayak, K., Fairbrother, D., & D'Souza, N. A. (2005). *Polymer, 46(23),* 7323–7333
4. Kozlov, G. V., & Miktaev, A. K. (2013). *Structure and Properties of Nanocomposites Polymer/organoclay,* LAP LAMBERT, Academic Publishing GmbH, Saarbrücken, 318.
5. Aharony, A. & Harris, A. B. (1989). *J. Stat. Phys., 54(3/4),* 1091–1097.
6. Vannimenus (1989). *J. Physical D, 38(2),* 351–355.
7. Kozlov, G. V., & Zaikov, G. E. (2004). *Structure of the Polymer Amorphous State.* Brill Academic Publishers, Utrecht, Boston, 465.
8. Balankin, A. S. (1991). *Synergetic of De formable Body,* Publishers of Ministry Defense of SSSR, Moscow, 404.
9. Kozlov, G. V., & Sanditov, D. S. (1994). *An harmonic Effects and Physical-Mechanical Properties of Polymers.* Nauka, Novosibirsk, 261.
10. Balankin, A. S., & Bugrimov, A. L. (1992). *Vysokomolek Soed A, 34(3),* 129–132.
11. Budtov, V. P. (1992). *Physical Chemistry of Polymer Solutions,* Khimiya, Sankt-Peterburg, 384.

CHAPTER 15

ACHIEVEMENTS IN DESIGN AND SYNTHESIS OF HYDROGEL-BASED SUPPORTS

D. HORÁK, AND H. HLHDKOVÁ

CONTENTS

ABSTRACT

Superporous poly (2-hydroxyethyl methacrylate) (PHEMA) supports with pore size from tens to hundreds micrometers were prepared by radical polymerization of 2-hydroxyethyl methacrylate (HEMA) with 2 wt.% ethylene dimethacrylate (EDMA) with the aim to obtain a support for cell cultivation. Super pores were created by the salt-leaching technique using NaCl or $(NH_4)_2SO_4$ as a porogen. Addition of liquid porogen (cyclohexanol/dodecan-1-ol (CyOH/DOH) = 9/1w/w) to the polymerization mixture did not considerably affect formation of meso and macro pores. The prepared scaffolds were characterized by several methods including water and cyclohexane regain by centrifugation, water regain by suction, scanning electron microscopy (SEM), mercury porosimetry and dynamic desorption of nitrogen. High-vacuum scanning electron microscopy (HVSEM) confirmed permeability of hydrogels to 8-μm microspheres, whereas low-vacuum scanning electron microscopy (LVSEM) at cryo-conditions showed the un-deformed structure of the frozen hydro gels. Interconnection of pores in the PHEMA scaffolds was proved. Water regain determined by centrifugation method did not include volume of large super pores (imprints of porogen crystals), in contrast to water regain by suction method. The porosities of the constructs ranging from 81 to 91% were proportional to the volume of porogen in the feed.

15.1   INTRODUCTION

Polymer supports have received much attention as microenvironment for cell adhesion, proliferation, migration and differentiation in tissue engineering and regenerative medicine. The three-dimensional scaffold structure provides support for high level of tissue organization and remodeling. Regeneration of different tissues, such as bone [1], cartilage [2], skin [3], nerves [4], or blood vessels [5] is investigated using such constructs. An ideal polymer scaffold should thus mimic the living tissue, that is, possess high water content, with possibility to incorporate bioactive molecules allowing a better control of cell differentiation. At the same time it requires a range of properties including biocompatibility and/or biodegradability, highly porous structure with communicating pores allowing high cell adhesion and tissue in-growth. The material should be sterilizable and also possess good mechanical strength. Both natural and synthetic hydrogels are being developed. The advantage of synthetic polymer matrices consists in their easy process ability, tunable physical and chemical properties, susceptibility to modifications and possibly controlled degradation.

Many techniques have been developed to fabricate highly porous constructs for tissue engineering. They include for instance solvent casting [6], gas foaming [7] or/ and salt leaching [8], freeze-thaw procedure [9, 10], supercritical fluid technology [11] (disks exposed to CO_2 at high pressure) and electrospinning (for nano fiber matrices) [12]. A wide range of polymers was suggested for scaffolds. In addition to natural materials, such as collagen, gelatin, dextran [13], chitosan [14], phosphorylcholine [15], alginic [16] and hyaluronic acids, it includes also synthetic polymers, e.g., poly(vinyl alcohol) [17], poly(lactic acid) [1, 18], polycaprolactone [19], poly(ethylene glycol) [20], polyacrylamide [21], polyphosphazenes [22], as well as polyurethane [23].

Among various kinds of materials being used in biomedical and pharmaceutical applications, hydrogels composed of hydrophilic polymers or copolymers find a unique place. They have a highly water-swollen rubbery three-dimensional structure, which is similar to natural tissue [24, 25]. In this report, poly (2-hydroxyethyl methacrylate) (PHEMA) was selected as a suitable hydrogel intended for cell cultivation. The presence of hydroxy and carboxy groups makes this polymer compatible with water, whereas the hydrophobic methyl groups and backbone impart hydrolytic stability to the polymer and support the mechanical strength of the matrix [26]. PHEMA hydrogels are known for their resistance to high temperatures, acid and alkaline hydrolysis and low reactivity with amines [27]. Previously, porous structure in PHEMA hydrogels was obtained by phase separation using a low molecular weight or polymeric porogen, or by the salt-leaching method. The material was used as a mouse embryonic stem cell support [8, 28, 29]. The aim of this report is to demonstrate conditions under which communicating pores are formed enabling high permeability of PHEMA scaffolds, which is crucial for future cell seeding.

15.2 EXPERIMENTAL PART

15.2.1 REAGENTS

2-Hydroxyethyl methacrylate (HEMA; Röhm GmbH, Germany) and ethylene dimethacrylate (EDMA; Ugilor S.A., France), were purified by distillation. 2,2'-Azobisisobutyronitrile (AIBN, Fluka) was crystallized from ethanol and used as initiator. Sodium chloride G. R. (Lach-Ner, s.r.o. Neratovice, Czech Republic) was classified, particle size 250–500μm and ammonium sulfate needles (100*600 μm, Lachema, Neratovice, Czech Republic) were used as porogens. Cyclohexanol (CyOH, Lachema, Neratovice, Czech Republic) was distilled, dodecan-1-ol (DOH) and all other solvents and reagents were obtained from Aldrich and used without purification. Ammonolyzed PGMA microspheres (2 μm) were obtained by the previously described procedure [30]. Sulfonated polystyrene (PSt) microspheres (8 μm) Ostion LG KS 0803 were purchased from Spolek pro chemickou a hutní výrobu, Ústí n. L., Czech Republic. Polyaniline hydrochloride microspheres (PANI, 200–400 nm) were prepared according to literature [31].

15.2.2   HYDROGEL PREPARATION

Cross-linked hydro gel constructs were prepared by the bulk radical polymerization of a reaction mixture containing monomer (HEMA), cross-linking agent (EDMA), initiator (AIBN) and NaCl or/and liquid diluents as a porogen (CyOH/DOH = 9/1w/w). The compositions of polymerization mixtures are summarized in Table 15.1. The amount of cross linker (2 wt. %) and AIBN (1wt %) dissolved in monomers was the same in all experiments, while the amount of porogen in the polymerization batch was varied from 35.9–41.4vol%. Optionally, needle-like $(NH_4)_2SO_4$ crystals (42.3vol %) together with saturated $(NH_4)_2SO_4$ solution were used as a porogen instead of NaCl crystals, allowing thus formation of hydro gels with communicating pores (Run 9, Table 15.1). For the sake of comparison, a copolymer was prepared with a mixture of solid (NaCl) and liquid low-molecular-weight porogen (diluents), amounting to 50% of the polymerization feed (Run 8). The thickness of the hydrogel was adjusted with a 3 mm-thick silicone rubber spacer between the Teflon plates (10 10 cm), greased with silicone oil and covered with Cellophane. The reaction mixture was transferred onto a hollow plate and covered with a second plate, clamped and heated at 70 °C for 8h. After polymerization, the hydrogels obtained with inorganic salt as a porogen were soaked in water and washed until the reaction of chloride or sulfate ions disappeared. The scaffolds prepared in the presence of a liquid diluent were washed with ethanol/water mixtures (98/2, 70/30, 40/60, 10/90 v/v) and water to remove the diluent, unreacted monomers and initiator residues. The washing water was then changed every day for 2 weeks.

TABLE 15.1   Preparation of PHEMA Hydrogels, Conditions and Properties[a]

Run	NaCl (vol.%)	Water regain (ml/g)		CX regain[d] (ml/g)	Cumulative pore volume[f] (ml/g)
1	41.4	0.84[d]	7.52[e]	0.13	0.35
2	40.8	0.88[d]	7.40[e]	0.23	0.47
3	40.0	1.04[d]	5.34[e]	0.56	1.03
4	39.1	0.81[d]	4.05[e]	0.33	1.24
5	37.9	0.78[d]	4.04[e]	0.21	1.60
6	37.0	0.79[d]	3.64[e]	0.34	1.83
7	35.9	0.75[d]	3.32[e]	0.32	1.70
8[b]	37.9	0.89[d]	4.30[e]	0.45	1.65
9	42.3[c]	0.84[d]	2.11[e]	0.08	0.08

[a] Crosslinked with 2 wt.% EDMA, 1 wt.% AIBN, NaCl in vol.% relative to polymerization mixture (HEMA + EDMA + NaCl).
[b] One half of the HEMA/EDMA feed was replaced by CyOH/DOH = 9/1 w/w.
[c] $(NH_4)_2SO_4$.
[d] Centrifugation method.
[e] Suction method.
[f] Mercury porosimetry.

15.2.3   METHODS

15.2.3.1   MICROSCOPY

LOW-VACUUM SCANNING ELECTRON MICROSCOPY

Low-vacuum scanning electron microscopy (LVSEM) was performed with a microscope Quanta 200 FEG (FEI, Czech Republic). Neat hydrated hydrogels were cut with a razor blade into ~ 5 mm cubes, flash-frozen in liquid nitrogen and placed on the sample stage cooled to −10 °C. Before microscopic observation, the top of a frozen sample was cut off using a sharp blade. During the observation, the conditions in the microscope (–10 °C, 100Pa) caused slow sublimation of ice from the sample, which made it possible to visualize its 3D morphology. All samples were observed with a low-vacuum secondary electron detector, using the accelerating voltage 30kV. Lyophilized PHEMA hydrogels Run 3 and Run 8 filled with microspheres were also examined by LVSEM; however, the microspheres were washed out of the pores during freezing and, consequently, they were scarcely observed on the micrographs.

HIGH-VACUUM SCANNING ELECTRON MICROSCOPY

High-vacuum scanning electron microscopy (HVSEM) was carried out with an electron microscope Vega TS 51355 (Tescan, Czech Republic). Permeability of the water-imbibed hydrogels (Runs 7 and 9) was investigated by the flow of water suspension of the polymer microspheres. Before observation, the wet hydrogel was placed on a wet filtration paper and a droplet of a suspension of 8-μm PSt microspheres in water was placed on the top. The sample was dried at ambient temperature and cut with a sharp blade in the direction of the microsphere flow. Samples showing the top, bottom and cross-sections of flowed-through hydrogels were sputtered with a 8-nm layer of platinum using a vacuum sputter coater (Baltec SCD 050), fixed with a conductive paste to a brass support and viewed in a scanning electron microscope in high vacuum (10^{-3} Pa), using the acceleration voltage 30kV and a secondary-electrons detector. This technique made it possible to observe both microspheres and pores of the hydrogel.

15.2.3.2   SOLVENT REGAIN

The solvent (water or cyclohexane-CX) regain was determined in 1×2 cm sponge pieces of hydrogel kept for 1 week in deionized water, which was exchanged daily. Water regain was measured by two methods: (i) centrifugation [32] (WR_c) and (ii) suction (WR_s). In centrifugation method, solvent-swollen samples were placed into glass columns with fritted disc, centrifuged at 980 g for 10 min and immediately weighed (w_w–weight of hydrated sample), then vacuum-dried at 80 °C for 7h and again weighed (w_d–weight of dry sample). In the second method, excessive water was removed from the imbibed hydrogel by suction and the hydrogel weighted to determine w_w. Weight of dry sample w_d was determined as above. Water regains WR_c or WR_s (ml/g) were calculated according to the equation:

$$WR_c \left(WR_s \right) = \frac{w_w - w_d}{w_d} \tag{1}$$

The results are average values of two measurements for each hydrogel. To measure cyclohexane regain (CXR) by centrifugation, equilibrium water-swollen hydrogels were successively washed with ethanol, acetone and finally cyclohexane. Using the solvent-exchange, a thermodynamically good (swelling) solvent in the swollen gel was replaced by a thermodynamically poor solvent (nonsolvent). Porosity of the hydrogels (p) was calculated from the water and cyclohexane regains (Table 15.1) and PHEMA density ($\rho = 1.3$ g/mL) according to the equation:

$$p = \frac{R \times 100}{R + \frac{1}{\rho}} \; (\%) \tag{2}$$

where $R = WR_c$, WR_s, or CXR (ml/g).

15.2.3.3   MERCURY POROSIMETRY

Pore structure of freeze-dried PHEMA scaffolds was characterized on a mercury porosimeter Pascal 140 and 440 (Thermo Finigan, Rodano, Italy). It works in two pressure intervals, 0–400kPa and 1–400 MPa, allowing determination of meso (2–50 nm), macro (50–1000 nm) and small superpores (1–116 µm). The pore volume and most frequent pore diameter were calculated under the assumption of a cylindrical pore model by the PASCAL program. It employed Washburn's equation describing capillary flow in porous materials [33]. The volumes of bottle and spherical pores were evaluated as the difference between the end values on the volume/pressure curve. Porosity was calculated according to Eq. 2, where cumulative pore volume (meso-, macro and small superpores) from mercury porosimetry was used for R.

15.3   RESULTS AND DISCUSSION

15.3.1   MORPHOLOGY OF HYDROGELS

The prepared PHEMA constructs had always an opaque appearance indicating a permanent porous structure. Pores are generally divided into micro-, meso-, macropores and small and large superpores. Morphology of water-swollen PHEMA hydrogels was investigated by LVSEM as shown in Fig. 15.1. Large 200–500 µm super pores were developed as imprints of NaCl crystals, which were subsequently washed out from the hydrogel; the interstitial space between them was filled with the polymer. During the observation, ice crystals filling soft polymer net were clearly visible in the center of the hydrogel Run 1 (Table 15.1) prepared at the highest content of NaCl (41.4 vol%) in the feed (Fig. 15.1a). The internal surface area was too small to be determined. Figures 15.1b and 15.1c show hydrogels from Run 3 and 5 (40 and 37.9 vol.% NaCl), respectively, documenting their more compact structure accompanied by thicker walls between large superpores as compared with the hydrogel from Run 1 (Fig. 15.1a). According to LVSEM, *ca.* 8-µm pores, the presence of which was confirmed by mercury porosimetry (volume about 1 mL/g), were observed in the walls between the large superpores. Longitudinal cracks in the material structure (Fig. 15.1b) were obviously caused by sample handling and fast freezing in liquid nitrogen. Nevertheless, the LVSEM micrographs displayed only cross-sections of hydrogels and it was not clear whether their pores are interconnected.

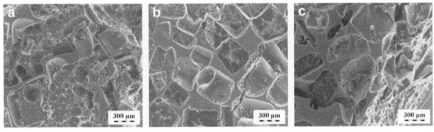

FIGURE 15.1   LVSEM micrographs showing frozen cross-section of PHEMA hydrogels prepared with (a) 41.4 vol.% – Run 1, (b) 40 vol.% – Run 3, and (c) 37.9 vol.% NaCl (250–500 µm)   Run 5. PHEMA cross-linked with 2 wt.% EDMA (relative to monomers).

Interconnection of pores is of vital importance for cell ingrowths in future applications to tissue regeneration. This feature was tested by the permeability of the whole hydrogels for different kinds of microspheres under two microscopic observations. First, cross-sections of the frozen hydrogels filled with microspheres were observed in LVSEM. Second, the water-swollen hydrogels were flowed through by a suspension of microspheres in water and their dried cross-sections were then viewed in HVSEM.

LVSEM showed water-swollen morphology of hydrogel constructs, which preserved due to their freezing in liquid nitrogen. PHEMA Run 3 prepared with neat NaCl (Fig. 15.2a) was compared with Run 8 obtained in the presence of NaCl together with a mixture of CyOH/DOH (Fig. 15.2d). Addition of liquid porogens did not change the morphology; however, it increased the pore volume (from 0.21 to 0.45 mL CX/g, Table 15.1) and softness of the hydrogel. As a result, it had a tendency to disintegrate during the washing procedure. LVSEM of both PHEMA hydrogels filled with 2-μm ammonolyzed PGMA and 200–400 nm PANI micro-spheres is illustrated in Figs. 15.2b, 15.2e and 15.2c, 15.2f, respectively. The mi-crographs showed undistorted morphology of the frozen hydrogels, but just a few microspheres and/or their agglomerates. This was attributed to the fact that most of them were washed out during preparation of the sample for LVSEM.

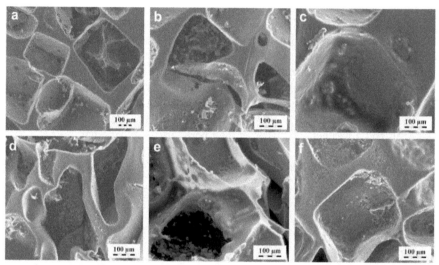

FIGURE 15.2   LVSEM micrographs showing frozen cross-section of PHEMA constructs; (a–c)-Run 3, (d–f)-Run 8; (a, d) neat and filled with (b, e) 2-μm ammonolyzed PGMA and (c, f) 200–400 nm PANI microspheres.

Morphology of the PHEMA hydrogels flowed through by a suspension of mi-crospheres was observed by HVSEM. Figure 15.3 shows HVSEM micrographs of cross-sections of the top and bottom part of the hydrogels from Runs 3, 5 and 8 flowed through by a suspension of 8-μm sulfonated PSt microspheres in water. While the microspheres flowed through the hydrogel construct from Run 3, they did not penetrate the ones from Run 5 and 8 prepared in the presence of a rather low content of NaCl (37.9 vol.%). At the same time, surface and inner structure of the hydrogels slightly differed. Figure 15.4 shows HVSEM of the longitudinal section of the PHEMA scaffold obtained with cubic NaCl crystals as a porogen (Run 7). While (Fig. 15.4) a shows the bulk, (Fig. 15.4b–d) detailed sections. Again, small superpores with an average size about 13 μm were in the walls between the large su-per pores, forming small channels through which water flowed. To prove or exclude the interconnection of at least some pores, a suspension of 8-μm sulfonated polysty-rene (PSt) microspheres in water was poured on the center of the topside of the gel. While water flowed through the hydrogel bulk, the microspheres were retained on the surface of the hydrogel or penetrated only superficial layers due to the surface cracks (Fig. 15.4a and 15.4b). This confirmed that the pores of PHEMA hydrogels obtained with a low content of NaCl porogen (35.9 vol. %) did not communicate. In contrast, Fig. 15.5 presents longitudinal section of the PHEMA hydrogel from Run 9 (both bulk and detailed) obtained with needle-like $(NH_4)_2SO_4$ crystals as a porogen. This porogen allowed formation of connected pores, which is explained by the needle-like structure of ammonium sulfate crystals that are linked to the gel structure. At the same time, the crystals grew to large structures due to the presence of saturated $(NH_4)_2SO_4$ solution in the feed. As a result, long interconnected large super pores channels – were formed. This is documented in Fig. 15.5a–d by the fact that suspension of 8-μm sulfonated PSt microspheres in water deposited in the center of the topside of the hydrogel flowed through. The captured microspheres are well visible in Fig. 15.5b–d. They accumulated at the places of pore narrowing; their majority, however, was found on the bottom part of the hydrogel. In such a way, the flow of water suspension of microspheres in the hydrogel was traced.

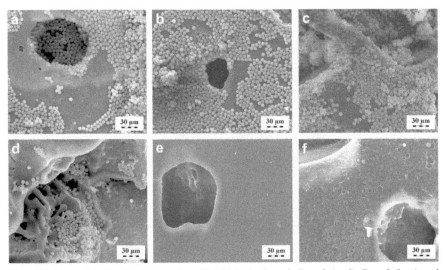

FIGURE 15.3 HVSEM micrographs of PHEMA hydrogels Run 3 (a, d), Run 5 (b, e) and Run 8 (c, f) showing top (a–c) and bottom (d–f) of the hydrogels after the flow of a suspension of 8-μm sulfonated PSt microspheres in water.

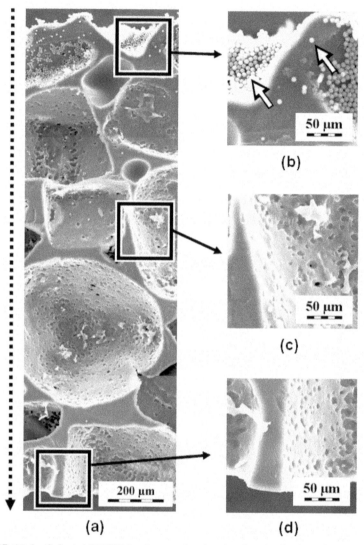

FIGURE 15.4   Selected HVSEM micrographs showing longitudinal section of PHEMA hydrogel 3 mm thick (Run 7) obtained with NaCl (250–500 μm) as a porogen after passing of a suspension of 8-μm sulfonated PSt microspheres in water (in the direction of the dotted line). (a) The whole cross-section through the hydrogel and selected details from (b) top, (c) center and (d) bottom. PSt microspheres are denoted with white arrows.

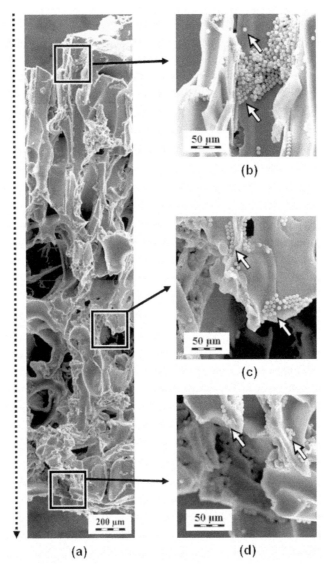

FIGURE 15.5 Selected HVSEM micrographs showing longitudinal section of PHEMA hydrogel 3 mm thick (Run 9) obtained with $(NH_4)_2SO_4$ (100×600 μm) as a porogen after passing of a suspension of 8-μm sulfonated PSt microspheres in water (in the direction of the dotted line). (a) The whole section through the construct and selected details from (b) top, (c) center and (d) bottom. PSt microspheres are denoted with white arrows.

Mechanical properties of the porous constructs were sensitive to the concentration of porogen in the feed. Hydrogels with lower contents of NaCl and therefore higher proportion of PHEMA had thicker walls between the pores and were more compact allowing increased swelling of polymer chains in water. Two PHEMA hydrogels with the highest contents of NaCl in the feed (41.4 vol.% – Run 1 and 40.8 vol.% – Run 2) possessing thin polymer walls between large superpores easily disintegrated as well as hydrogel prepared using $(NH_4)_2SO_4$ (42.3 vol.% – Run 9).

15.3.2   CHARACTERIZATION OF POROSITY BY SOLVENT REGAIN

Dependences of porosity of PHEMA hydrogels calculated from water or cyclohexane regain and also from mercury porosimetry on NaCl content in the polymerization feed showed similar behavior (Fig. 15.6). Porosities 81–91 and 49–57% for water regain were obtained by suction and centrifugation, respectively, 14–42 % for cyclohexane regain and 31–70% for mercury porosimetry. The porosity determined by centrifugation of samples soaked with water and cyclohexane (solvents with different affinities to polar methacrylate chain) consists of two contributions: filling of the pores and swelling (solvation) of PHEMA chains. The uptake of cyclohexane, a thermodynamically poor solvent that cannot swell the polymer, is a result of the former contribution only, reflecting thus the pore volume. The water regain from centrifugation was always higher than the cyclohexane regain demonstrating thus swelling of polymer chains with water (Table 15.1). Solvent regains were affected by the concentration of NaCl porogen in the polymerization feed. Porosities according to both water and cyclohexane regains by centrifugation slightly increased with increasing volume of NaCl porogen in the polymerization feed from 35.9 to 40 vol.% and then decreased with a further NaCl increase up to 41.4 vol.% (Fig. 15.6). In the latter range of NaCl, the porosity evaluated by mercury porosimetry exhibited an analogous dependence. This decrease in solvent and mercury regains can be explained by thin polymer walls between large superpores inducing collapse of the porous structure. In the concentration range of NaCl in the feed 35.9–40 vol. %, mercury porosimetry provided higher porosities than those obtained from regains by centrifugation at 980 g because it obviously did not retain solvents in large super pores. Retained water reflected thus only small superpores, closed pores and solvation of the polymer in water similarly as observed earlier for macroporous PHEMA scaffolds [34]. Water regain was determined also by the suction method (Table 15.1) which gave the values several times higher (3.3–7.5 mL/g) than by centrifugation due to filling all the pores in the polymer structure, including large superpores. As expected, porosity by the suction method increased with increasing volume of NaCl porogen in the polymerization feed (Fig. 15.6).

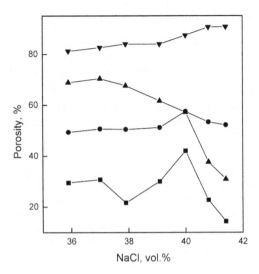

FIGURE 15.6 Dependence of porosity of PHEMA hydrogels determined from cyclohexane (■) and water regain measured by centrifugation (●) or suction (▼) and mercury porosimetry (▲) on the content of NaCl (250–500 μm) porogen in the polymerization feed.

The hydrogel from Run 8 formed in the presence of NaCl and CyOH/DOH porogen showed higher solvent regains and mercury penetration than the comparable hydrogel from Run 5 obtained with the same content of neat NaCl (Table 15.1). This can be explained by the higher total amount of porogen in the former hydrogel. In contrast, the hydrogel from Run 9 prepared with needle-like $(NH_4)_2SO_4$ crystals as a porogen had the lowest solvent and mercury regains of all the samples. The exception was water regain by centrifugation, which was identical with that of sample Run 1 (Table 15.1) having a similar content of the NaCl porogen in the feed. This can imply that only large continuous superpores were present in this hydrogel and small superpores, macro and mesopore were almost absent as evidenced by the low values of solvent and mercury regains.

15.3.3 CHARACTERIZATION OF POROSITY BY MERCURY POROSIMETRY

The advantage of mercury porosimetry is that it provides not only pore volumes, but also pore size distribution not available by other techniques. The method measures samples dried by lyophilization, which does not distort the pore structure. As already mentioned, porosities determined by mercury porosimetry were lower than those obtained from water regain by the suction, which included large superpores, and higher than those from water and cyclohexane regain detected by centrifugation. This was due to better filling of the compact xerogel structures obtained at lower contents of NaCl in the feed with mercury under a high pressure than with water or cyclohexane under atmospheric pressure. Figure 15.7 shows the dependence of most frequent mesopore size of PHEMA scaffolds and their pore volumes on the NaCl porogen content in the polymerization feed. Predominantly, 4–5 nm mesopores were detected with their volume increasing from 0.03 to 0.1 mL/g with increasing NaCl content in the polymerization feed. Macropores were absent and very low values of specific surface areas (<0.1 m^2/g) were found. The presence of CyOH/DOH porogen (Run 8) did not substantially affect the formation of meso- and macropores (volume 0.022 mL/g), because the amount of cross-linker in the polymerization feed was limited to only 2 wt.%. The separation of the polymer from the porogen phase could not thus occur and porous structure was not formed. Both in hydrophobic styrene-divinylbenzene [35, 36] and polar methacrylate copolymers [37] prepared in the presence of liquid porogens, phase separation and formation of macroporous structure occurred at cross-linker contents higher than 10 wt.%.

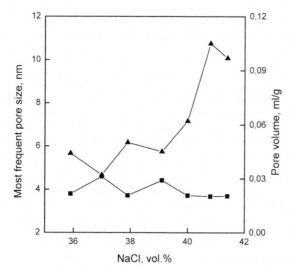

FIGURE 15.7   Dependence of pore volume (▲) and most frequent mesopore size (■) of PHEMA hydrogels on the content of NaCl (250–500 μm) porogen in the polymerization feed according to mercury porosimetry.

Figure 15.8 represents the dependence of most frequent small superpore size of PHEMA constructs and their pore volume on the content of NaCl porogen in the polymerization. Pore size increased up to 28–69 μm with increasing NaCl volume. This was pronounced in the range 40–41.4 vol.% NaCl probably due to the aggregation of NaCl crystals in the mixture at their high contents. All the in-vestigated samples contained small superpores, the volume of which was about 20 times higher than that of mesopores. The volume of small superpores continuously decreased from 1.8 to 0.2 mL/g with raising NaCl amount in the feed. This could be explained by collapse of the pore structure and destruction of last two hydrogels with the highest content of NaCl porogen (Runs 1 and 2) under high pressure as mentioned above. The size of small superpores according to mercury porosimetry was by an order of magnitude smaller than the particle size of the used NaCl poro-gen (250–500 μm) because the method was able to distinguish the superpores only in the size range 1–116 μm (Fig. 15.9). Large superpores (imprints of NaCl crystals) were detected by LVSEM. Figure 15.9 exemplifies a typical cumulative pore vol-ume and a derivative pore size distribution curve of PHEMA Run 4 with a decisive contribution of small superpores 25 μm in size.

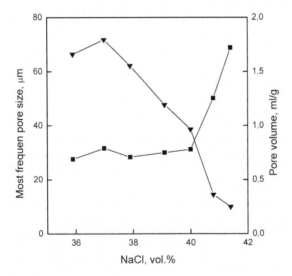

FIGURE 15.8 Dependence of pore volume (▼) and most frequent small superpore size (■) of PHEMA hydrogels on the content of NaCl (250–500 μm) porogen in the polymerization feed according to mercury porosimetry.

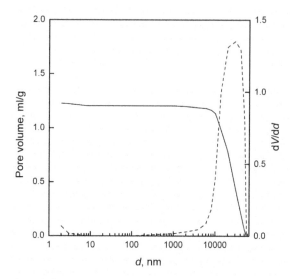

FIGURE 15.9   Cumulative pore volume (—) and pore size distribution (—) of the Run 4 hydrogel determined by mercury porosimetry in the range 1.88 nm – 116 μm.

15.4   CONCLUSIONS

Superporous PHEMA constructs were prepared by bulk radical copolymerization of HEMA and EDMA in the presence of NaCl or/and liquid diluent (CyOH/DOH) or $(NH_4)_2SO_4$ crystals. Morphology of the prepared scaffolds was characterized by several methods including scanning electron microscopy both in swollen (LVSEM) and dry (HVSEM) state, solvent (water and cyclohexane) regains, high- and low-pressure mercury porosimetry of lyophilized samples and dynamic desorption of nitrogen. Morphology and porous structure of the hydrogels were preferentially affected by the character and amount of the used porogen–NaCl, CyOH/DOH mixture or $(NH_4)_2SO_4$. After washing out of the salts and solvents from PHEMA, three types of pores were detected by microscopic and mercury porosimetry methods, including large superpores (hundreds of micrometers) as imprints of salt crystals. The hydrogels formed can be divided into two groups, with disconnected and interconnected pores. The latter allowed the passage of suspension of microspheres in water, which was observed only for the samples with ammonium sulfate and the highest content of NaCl crystals used as a porogen in the feed. Interconnected pores are crucial for potential application of the scaffolds as living cell supports. LVSEM showed the undistorted (frozen) structure of the hydrogels, but only few flowed-through microspheres could be observed as they tended to escape from the pores during sample preparation. HVSEM seemed to be the best microscopic technique especially for viewing permeability of hydrogels to 8-μm microspheres. Hydrogels were initially flowed through by the particles in their natural wet state, but the specimens were then dried before SEM observation. The microparticles could be traced both on the upper/lower parts of the hydrogels and on the longitudinal sections.

Mercury porosimetry provided detailed description of morphology of PHEMA constructs with pore sizes from units of nanometers to tens of micrometers. The drawback of the method is that the hydrogels are not measured in the swollen, but dry state, as xerogels. But comparison of the data in both wet and dry states showed that lyophilization did not change the pore structure. The mesopores and small superpores detected by mercury porosimetry cannot be formed by the imprinting mechanism. While mesopores present only in very small amounts may be formed by phase separation, small superpores arise by polymer contraction in the walls of large superpores. Small 28–69μm super pores were mainly present in the porous structure apart from the large super pores (200–500 μm imprints of solid porogen crystals), the volume of which was several times higher than that of other pores, as confirmed by water regain obtained by the suction method.

ACKNOWLEDGMENT

Financial support of the Grant Agency of the Czech Republic (project No. P304/11/0731) is gratefully acknowledged.

KEYWORDS

- 2-Hydroxyethyl Methacrylate
- Hydrogel
- Porosity
- Scaffold

REFERENCES

1. Kofron, M. D., Cooper, J. A., Kumbar, S. G., & Laurencin, C. T. (2007). Novel Tubular Composite Matrix for bone repair. *J Biomed Mater Res, Part A; 82*, 415–425.
2. Moroni, L., Hendriks, JA. A., Schotel, R., De Wijn, J. R., & Van Blitterswijk, C. A. (2007). Design of biphasic polymeric 3-dimensional fiber deposited scaffolds for cartilage tissue engineering applications. *Tissue Eng; 13*, 361–371.
3. Dvořánková, B., Holíková, Z., Vacík, J., Königová, R., Kapounková, Z., Michálek, J., Přádný, M., & Smetana, K. (2003). Reconstruction of epidermis by grafting of keratinocytes cultured on polymer support clinical study. *Int J Dermatol*; *42*, 219–223.
4. Bhang, S. H., Lim, J. S., Choi, C. Y., Kwon, Y. K., & Kim, B. S. (2007). The Behavior of Neural Stem cells on Biodegradable Synthetic Polymers. *J Biomater. Sci., Polym 18*, 223–239.

5. Bianchi, F., Vassalle, C., Simonetti, M., Vozzi, G., Domenici, C., & Ahluwalia, A. (2006). Endothelial cell function on 2D and 3D microfabricated polymer scaffolds: Applications in cardiovascular tissue engineering. *J Biomater Sci., Polym 17*, 37–51.

6. Sander, E. A., Alb, A. M., Nauman, E. A., Reed, W. F., & Dee, K. C. (2004). Solvent effects on the microstructure and properties of 75/25 poly(D, L-lactide-coglycolide) tissue scaffolds. *J Biomed Mater Res, Part A, 70*, 506–513.

7. Nam, Y. S., Yoon, J. J., & Park, T. G. (2000). A novel fabrication method of macroporous biodegradable polymer scaffolds using gas foaming salt as a porogen additive. *J Biomed Mater Res, Appl Biomater; 53*, 1–7.

8. Kroupová, J., Horák, D., Pacherník, J., Dvořák, P., & Šlouf, M. (2006). Functional polymer hydrogels for embryonic stem cell support. *J Biomed Mater Res, Part B: Appl Biomater 76B*, 315–325.

9. Tighe, B., & Corkhill, P. (1994). Hydrogels in biomaterials design: Is there life after polyHEMA? *Macromol Rep; A31*, 707–713.

10. Plieva, F. M., Galaev, I. Y., & Mattiasson, B. (2007). Macroporous gels prepared at subzero temperatures as novel materials for chromatography of particulate-containing fluids and cell culture applications. *J Sep Sci; 30*, 1657–1671.

11. Mooney, D. J., Baldwin, D. F., Suh, N. P., Vacanti, J. P., & Langer, R. (1996). Novel approach to fabricate porous sponges of poly(D, L-lactic-coglycolic acid) without the use of organic solvents. *Biomaterials; 17*, 1417–1422.

12. Chung, H. J., & Park, T. G. (2007). Surface engineered and drug releasing prefabricated scaffolds for tissue engineering. *Adv Drug Delivery Rev; 59*, 249–262.

13. Ferreira, L. S., Gerecht, S., Fuller, J., Shieh, H. F., Vunjak-Novakovic, G., & Langer, R. (2007). Bioactive hydrogel scaffolds for controllable vascular differentiation of human embryonic stem cells. *Biomaterials; 28*, 2706–2717.

14. Tangsadthakun, C., Kanokpanont, S., Sanchavanakit, N., Pichyangkura, R., Banaprasert, T., Tabata, Y., & Damrongsakkul, S. (2007). The influence of molecular weight of chitosan on the physical and biological properties of collagen/chitosan scaffolds. *J Biomater Sci, Polym 18*, 147–163.

15. Wachiralarpphaithoon, C., Iwasaki, Y., & Akiyoshi, K. (2007). Enzyme-degradable phosphorylcholine porous hydrogels cross-linked with polyphosphoesters for cell matrices. *Biomaterials, 28*, 984–993.

16. Treml, H., Woelki, S., & Kohler, H. H. (2003). Theory of capillary formation in alginate gels. *Chem Phys, 293*, 341–353.

17. Konno, T., & Ishihara, K. (2007). Temporal and spatially controllable cell encapsulation using a water-soluble phospholipid polymer with phenylboronic acid moiety. *Biomaterials, 28*, 1770–1777.

18. Heckmann, L., Schlenker, H. J., Fiedler, J., Brenner, R., Dauner, M., Bergenthal, G., Mattes, T., Claes, L., & Ignatius, A. (2007). Human mesenchymal progenitor cell responses to a novel textured poly(L-lactide) scaffold for ligament tissue engineering. *J Biomed Mater Res, Part B, Appl Biomaterial, 81*, 82–90.

19. Darling, A. L., & Sun, W. (2004). 3D Microtomographic Characterization of precision extruded poly ε-Caprolactone scaffolds. *J Biomed Mater Res, Part B, Appl Biomaterial, 70*, 311–317.

20. Rhee, W., Rosenblatt, J., Castro, M., Schroeder, J., Rao, P. R., Harner, C. F. H., & Berg, R. A. (1997). In vivo stability of poly(ethylene glycol)-collagen composites, in:

Poly(Ethylene Glycol) *Chemistry and Biological Applications*, Harris, J. M., Zalipsky, S., Eds, *ACS Symp Ser; 680*, 420–440.

21. Savina, I. N., Galaev, I. Y., & Mattiasson, B. (2006). Ion-exchange macroporous hydrophilic gel monolith with grafted polymer brushes. *J Mol Recognit; 19*, 313–321.

22. Carampin, P., Conconi, M. T., Lora, S., Menti, A. M., Baiguera, S., Bellini, S., Grandi, C., & Parnigotto, P. P. (2007). Electrospun polyphosphazene nanofibers for in vitro rat endothelial cells proliferation. *J Biomed Mater Res, Part A; 80*, 661–668.

23. Zhang, C. H., Zhang, N., & Wen, X. J. (2007). Synthesis and Characterization of Biocompatible, degradable, light curable, polyurethane based elastic hydro gels. *J Biomed Mater Res, Part A; 82*, 637–650.

24. Castner, D. G., & Ratner, B. D. (2002). Biomedical surface science: Foundation to frontiers. *Surf Sci, 500*, 28–60.

25. Lee, K. Y., & Mooney, D. J. (2001). Hydro gels for tissue engineering, *Chem. Rev, 101*, 1869–1879.

26. Refojo, M. F. (1967). Hydrophobic interactions in poly(2-hydroxyethyl methacrylate) homogeneous hydrogel. *J Polym Sci, Part A1, Polym Chem, 5*, 3103–3108.

27. Ratner, B. D., & Hoffman, A. S. (1976). Hydrogels for Medical and Related Applications, *ACS Symp Ser, 31*, 1–36.

28. Horák, D., Dvořák, P., Hampl, A., & Šlouf M. (2003). Poly(2-hydroxyethyl methacrylate-co-ethylene dimethacrylate) as a mouse embryonic stem cell support, *J Appl Polym Sci, 87*, 425–432.

29. Horák, D., Kroupová, J., Šlouf, M., & Dvořák P. (2004). Poly(2-hydroxyethyl methacrylate) based slabs as a mouse embryonic stem cell support. *Biomaterials, 25*, 5249–5260.

30. Horák, D., & Shapoval, P. (2000). Reactive poly(glycidyl methacrylate) microspheres prepared by dispersion polymerization. *J Polym Sci, Part A, Polym Chem Ed, 38*, 3855–3863.

31. Stejskal, J., Kratochvíl, P., Gospodinova, N., & Terlemezyan, L., & Mokreva, P. (1992). Polyaniline dispersions: Preparation of spherical particles and their light-scattering characterization. *Polymer, 33*, 4857–4858.

32. Štamberg, J., & Ševčík S. (1966). Chemical transformations of polymers III. Selective hydrolysis of a copolymer of diethylene glycol methacrylate and diethylene glycol dimethacrylate. *Collect Czech Chem Commun; 31*, 1009–2016.

33. Porosimeter Pascal 140 and Pascal 440, Instruction manual, 8.

34. Hradil, J., & Horák D. (2005). Characterization of pore structure of PHEMA based slabs, *React Funct Polym., 62*, 1–9.

35. Millar, J. A., Smith, D. G., Marr, W. E., & Kresmann, T. R. E. (1963). Solvent modified polymer networks. Part 1. The preparation and characterization of expanded networks and macroporous styrene-DVB copolymers and their sulfonates. *J Chem Soc*, 218–225.

36. Kun, K. A., & Kunin, R. (1968). Macroreticular resins III. Formation of Macro reticular Styrene-Divinyl benzene copolymers. *J Polym Sci, Part A1, Polym Chem. 6*, 2689–2701.

37. Hradil, J., Křiváková, M., Starý, P., & Čoupek, J. (1973). Chromatographic properties of macroporous copolymers of 2-hydroxyethyl methacrylate and ethylene dimethacrylate. *J Chromatogr. 79*, 99–105.

CHAPTER 16

STRUCTURAL CHANGES IN SYNTHETIC MINERALS

A. M. IGNATOVA and M. N. IGNATOV

CONTENTS

ABSTRACT

This study investigates the structural changes in synthetic minerals obtained by casting under shock impact, specifically, the characteristic changes in each mineral phase.

16.1   INTRODUCTION

It is known that many rocks are formed under intense impact effects (e.g., impact of meteorites). Deposits of minerals, including rare metals and nonmetal structures, are found in impact craters (astrobleme) [1–3]. Additionally, impact metamorphism forms polymorphs of certain minerals that cannot be formed under other conditions, such as the modification of quartz to form coesite and stishovite [4].

It has been established experimentally that some impact breeds can be synthesized. Such synthesis is possible under the impact of a small number of nonaltered rock hits under the condition that the force of the impact and volume of rock is well correlated with the force of meteorite impact and volume of mountain formation [5–6].

Clearly, impact effects change the structure of synthetic materials, not only naturally occurring materials. Consequently, impact can be considered a method of synthesis, not only a destructive phenomenon.

The principles of impact metamorphism are used, in part, for mechanical activation [7]. Indeed, the mechanical activation of powder materials can be used to obtain a variety of materials with very fine structure. However, the structural changes induced by mechanical activation are different from those induced by impact metamorphism. Shock metamorphism occurs due to the effects of impact on a monolithic object, while mechanical activation acts on powder materials. Thus, it is important to study the structural changes in materials in the solid and monolithic state under shock impact.

It is known that under shock impact, impact energy is transferred; this phenomenon is called energy dissipation. In this process, which has several different stages, the impact energy is converted to heat [8].

Different materials dissipate energy differently, depending on their structure. A good example is how materials behave when hit by a bullet from a gun. If a material is destroyed by a bullet that stops moving within the material, then the kinetic energy of the bullet is spent on the destruction and changes in the material. Conversely, if the bullet continues to move, and the material is damaged slightly, then such a material is not prone to dissipation. The more a material able to dissipate energy, the more its structure changes under shock impact. Rocks are materials with a high capacity for dissipation, as confirmed by the phenomenon of impact metamorphism described previously. Materials with structures similar to those of natural mineral formations will have a similar capacity for dissipation.

Rock materials such as glasses and ceramics are often used to produce armor. Armor is necessary not only for protection from firearms but also to protect turbine engine blades, for example, in the case of emergency isolation.

However, the materials with greatest similarity to natural mineral formations are synthetic mineral alloys (siminals). The term "siminals" is new, and it is intended to replace the outdated term "casting a stone."

Siminals are materials obtained by remelting one or more varieties of rocks or man-made mineral formations. Their structure and composition are similar to those of mafic and ultramafic igneous rocks. The structure contains a siminal amorphous phase (2 to 30%) and two or more mineral phases. Siminals are not glass-crystalline materials because they are not crystallized using catalysts [9]. Siminals contain an average of 50% SiO_2; therefore, siminal melts often split into two liquid phases. The structure of siminals represents a special case of a condensed medium.

16.2 AIM AND OBJECT OF STUDY

The aim of this study was to study structural changes in siminals, specifically raw hornblendite materials, under shock impact.

The chemical composition of the samples, as presented in Table 16.1, was obtained by X-ray spectral fluorescence (CPM-18, EDX 900HS).

TABLE 16.1 Composition of Samples Studied

The content of components, %								
MgO	Al_2O_3	SiO_2	K_2O	CaO	TiO_2	Cr	MnO	Fe_2O_3
9.91	10.82	49.6	0.22	10.33	1.38	0.54	0.16	14.02

16.3 RESEARCH METHODS AND RESULTS

High-speed penetration tests were conducted by subjecting samples to shock cylindrical punching at speeds of 125, 600 and 800 m/s [13]. At the moment of impact, the temperature distribution over the punch pattern on the backside of the samples was recorded using an infrared camera.

Changes in the structure of the samples were evaluated using X-ray diffraction and scanning electron microscopy, with associated spectral microprobe analysis. The fragments of the destroyed samples were studied.

X-ray diffraction was conducted using an ESR70–30 DX/ 2 diffractometer. Scanning electron microscopy and microprobe analysis were performed on a JSM-6390LV.

Radiographs of the samples after run off at a speed of 125 m/s are shown in Fig. 16.1. The spectrogram revealed that the main phase of the samples was clinopyroxene (diopside-augite composition), with olivine, quartz, and traces of chromite impurities also observed. The greatest interest in terms of the changes in the structure lies in the intensity of the peak at 2.96, which corresponds to one of the shock-induced polymorphs of quartz stishovite.

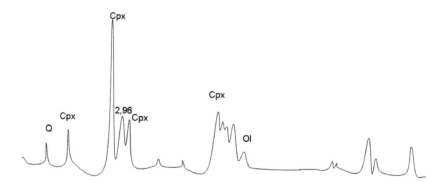

FIGURE 16.1   Radiographs of a specimen subjected to a shock speed of 125 m/s.

X-ray analysis of the siminal sample destroyed at a speed of 650 m/s (Fig. 16.2) revealed that the stishovite peak increased. This is indirect evidence of impact metamorphism. Meanwhile, the intensities of the rest of the peaks also increased, which corresponds to the appearance of impact-induced changes in pyroxene.

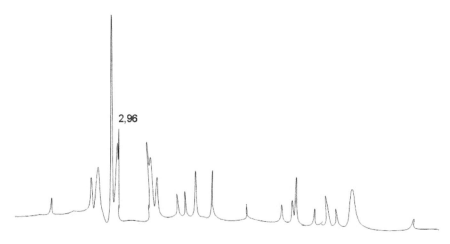

FIGURE 16.2   Spectrogram corresponding to the sample destroyed at an impact speed of 650 m/s.

Fragments obtained after the destruction of the targets were selected for sample analysis by scanning electron microscopy. Figure 16.3 shows images of the surface fragments of the sample destroyed at a punch speed of 125 m/s.

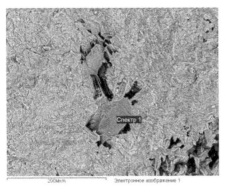 

FIGURE 16.3   Surface of fragments of the sample destroyed at a punch speed of 125 m/s.

The surfaces of the sample fragments show two zones: one has a pronounced relief surface, and the other features a smooth, kinked surface (this is typical for amorphous materials). Areas with different morphologies are separated by a small gap, which is likely generated during the process of destruction between the different phases of the cracks. The chemical composition of each area on the surface was determined. It was determined that the area showing "smooth" fracture is composed of SiO_2 (the rest of MgO, FeO); a crystalline fracture in the same area also contains SiO_2 but in much smaller quantities, with albite, magnesite, bauxite, wollastonite, feldspar, and also traces of iron oxide and titanium detected. The crystalline area is heterogeneous; thus, it was studied in greater detail at high magnification, which revealed that the morphology of the surface profile of the first projections of relief coincides with that induced by natural crystal growth. This result confirms that the initial crack was formed directly at the interface between two phases with different degrees of ordering.

Relief projections of the same shape and height on the tops of ridges are areas that differ from the rest of the material (Fig. 16.4). Spectral analysis revealed that the areas on the tops of ridges are composed of FeO, MgO and Al_2O_3, and Cr_2O_3 as well as 6–7% wollastonite, titanium oxide and silica in small quantities. The protrusions are composed of SiO_2 and high quantities of MgO and CaO, with the other components present in the same proportions; no chromium was detected.

The differences in the fracture morphology and composition of selected areas on the surface of samples broken at high speeds were considered.

FIGURE 16.4   Images showing the increase in the surface area of fragments of the sample destroyed at a punch speed of 125 m/s.

The fracture surface of the sample subjected to an impact speed of 650 m/s revealed two distinct zones: one, as observed previously, showed a morphology similar to that induced by natural crystal growth, while the other was smooth (Fig. 16.5a). The cracks were more extensive and represented the contours of crystalline aggregates. A clear shift in some areas relative to others was observed. The smooth sections show heterogeneity, and the crystalline zone features relief structures that show smaller differences in height but look similar to the projections.

Spectral analysis revealed that the smooth surfaces are composed of SiO_2 and MgO, as well as wollastonite and the oxides MnO and FeO. A characteristic feature is the presence of manganese oxide, which was not detected in samples obtained after collision at a slower rate. The surface morphology of the zone observed at high magnification (Fig. 16.5b) also revealed clear differences; the slip bands have relief, similar to the fracture surface observed within the same field in the first case, but at high velocities, the formation of wollastonite was observed.

The sample surface showing a crystalline morphology (Fig. 16.5c) is different in composition, with selected chlorine compounds found in small quantities. A more detailed study of the topography of the crystalline area revealed that the height difference was smaller in some areas. The surface is not acutely angled, and the crystallite size is generally smaller than that in the first case. The composition did not change.

The most widespread changes in the structure, surface morphology, and composition of the individual zones were detected in samples obtained by punching at the maximum speed of 800 m/s. First of all, there were no severe fractures; instead, cracking was observed at the folds (all fractures resulted in destruction). Three distinct areas were identified: dark areas of indeterminate morphology (Fig.16.6), crystalline areas of relief, which were much lower in quantity and smaller, and areas indicating crystallites on the tips of relief structures, many of which were markedly large in size. A study of the individual sections of the surface revealed that the dark areas contain carbon (which was not present before), which is important because carbon forms the basis of the impurity oxides occurring in the material, as well as small amounts of sulfur. Microprobe analysis indicates that these dark patches are most likely thin films.

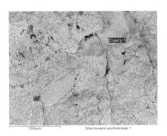

a

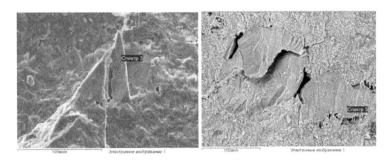

b

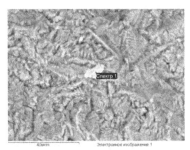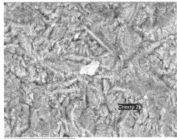

FIGURE 16.5   Surface of a fragment of the sample destroyed at a punch speed of 650 m/s: a – general view, b – a zone with slip bands, c – crystalline fracture zone.

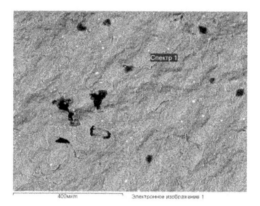

FIGURE 16.6 Surface of a fragment of the sample destroyed at a speed of 800 m/s; allocation of carbon on the fracture surface.

Studying the rest of the components at high magnification revealed that the shape of the crystallites containing chromium changed (Fig. 16.7). The formed mineral was first decomposed into its components – the skeleton of the crystal. Then, the mineral was deformed and its composition changed; the admixture of vanadium was detected, which was previously undetected. The crystalline mass over much of the surface leveled off at a certain height. Clearly delineating the boundaries of folds surrounding areas of fracture, the same crystallites are clearly distorted and fragmented, despite the fact that their base composition still contains silicon dioxide, the content of which largely increased; the concentration of sodium oxide and chromium oxide increased as a result of the fragmentation of the other phases.

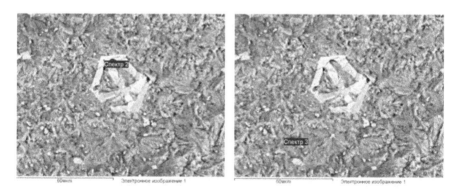

FIGURE 16.7 Destruction and deformation of structural components of siminals with penetration at speeds of 850 m/s.

16.4   DISCUSSION OF THE RESULTS

The results of X-ray diffraction, electron microscopy and microprobe studies suggest that the phenomenon of shock metamorphism occurred in the siminals. However, because the composition of the siminals is a synthetic analog of naturally occurring, shock-induced minerals, the changes consist of those in pyroxene, not in the polymorphic transformations, and the removal and redistribution of elements between the components because pyroxene is resistant to impact.

The pyroxene was composed principally of sodium and calcium, followed by silicon dioxide because of its ability to be transformed into stishovite. When undergoing a polymorphic transition to stishovite, silicon dioxide increases in volume and therefore can be quickly destroyed. It is important to establish why some elements were not detected in the samples before impact, particularly carbon. To understand where it came from, the process through which siminals are obtained should be analyzed.

The raw material used to produce siminals is coal melted by a graphite arc system. The graphite electrodes in the arc system react with the melt such that some carbon goes into the melt, as it is well dissolved. The interaction between the melt and carbon forms a supersaturated solid solution of carbon in clinopyroxene. A shock load applied to pyroxenes typically removes different elements, in particular, sodium, potassium and calcium, and forges interactions between the siminal material and carbon as well. This explains the presence of vanadium in several phases.

This information allows us to connect all of the changes in the dissipation of mechanical energy in siminal structures in the following sequence:

Shock $\rightarrow$ polymorphic transformation of quartz into stishovite (swelling of quartz) $\rightarrow$ local heating and pressure upon the accumulation of clinopyroxene expanding stishovite $\rightarrow$ Partial disordering of clinopyroxene $\rightarrow$ allocation of sodium and calcium $\rightarrow$ allocation of carbon, vanadium, sulfur and other elements of the solid solution and the diffusion of chromium $\rightarrow$ damage to the fine dust particle melt.

It is known that shock effects in materials produce elastic waves, contributing to the movement of structural defects (such as dislocations). The defects are consolidated; thus, during heat transfer, the vibrations of the surrounding atoms are enhanced. Therefore, during impact, the first phase to suffer the effects is that which exhibits a low degree of long-range order, that is, the amorphous phase. Low-temperature siminals typically feature quartz (a-quartz), which forms trigonal-trapezoidal crystals. The characteristic structural units of silica tetrahedra are connected to each other through their vertices. The tetrahedra form nonplanar six membered rings in the quartz crystal lattice. In amorphous silica, the tetrahedra are the main structural elements, but there are also eight-membered rings, with the planes of the rings rotated at different angles.

In quartz, the main chemical bond is Si-O. They are approximately equal amounts of homopolar and hetero polar bonds. Thus, the notion that the quartz crystal lattice is composed of Si and O atoms, or Si_4-O_2 ions, reflects only the limiting cases of real chemical bonds. This suggests the possibility of the homolytic and heterolytic rupture of Si-O upon impact. Quartz in all of its states is a fragile material; thus, upon impact it usually collapses. From mechanosynthesis, it is known that the destruction of quartz due to mechanical shock is similar to its dissolution or melting. According to the phase diagram of quartz, under a sharp blow, the probability of the conversion of quartz to stishovite is enhanced with the decreasing degree of crystallinity. Stishovite has a lower density than the low-temperature version of quartz. Upon impact, the vibrations of elastic waves resonate with the lattice vibrations of quartz. Due to this instantaneous pressure, quartz "swells," which induces stresses arising from relaxation by the formation of cracks. However, not all of the energy goes to the formation of cracks, which leads to heating.

When silicon oxide undergoes a polymorphic transformation, associated the pressure and temperature induce the formation of clinopyroxene in microvolumes. A sufficiently high-pressure force on crystallites of clinopyroxene will lead to intense amorphization. When Na and Ca2 are amorphised, they are pushed out of the matrix, and the effect of temperature will lead to the early melting of microparticles, initiating the decomposition of aluminosilicate tetrahedra.

This process describes the effects of impact on the chromium phase distributed in clinopyroxene. In this process, clinopyroxene is deformed and partially decomposed. Increasing the intensity of exposure leads to transformations in clinopyroxene (which forms a solid solution with carbon). As a result, pure carbon is allocated to the periphery of the crystallites.

16.5   CONCLUSIONS

The results allow us to assert that the effects of impact on siminals have some promise.

First, the effects of impact help remove alkaline impurities. By further remelting powders after impact, better and cleaner siminals can be obtained.

Second, by submitting them to high levels of impact, particles of chromite can be separated for further use as a high-hardness material.

Third, the allocation of carbon in the form of a separate phase suggests that large changes in velocity and pressure conditions can be induced to produce different polymorphs of carbon by the shock metamorphism of siminals.

Thus, the studied changes in the structure of siminals under shock impact form a consistent picture of the materials' behavior. Theoretically, impact could be used as a method of processing siminals to produce unique products.

KEYWORDS

- **Dissipation**
- **Impact mineral**
- **Shock metamorphism**
- **Synthetic mineral alloys**

REFERENCES

1. www.unb.ca/passc/ImpactDatabase Website describing 174 meteorite impacts world-wide. Developed and maintained by Planetary and Space Science Centre, University of New Brunswick, Fredericton, New Brunswick, Canada.
2. Collins, G. S., Melosh, J. H., & Marcus, R. A. (2005). Earth impact effects program: A web-based computer program for calculating the regional environmental consequences of a meteoroid impact on Earth; Meteorite and Planetary Science *40*: 817–840. (www.lpl.arizona.edu/impacteffects).
3. Addison, W. D., Brumpton, G. R., Vallini, D. A., McNaughton, N. J., Davis, D. W., Kissin, S. A., Fralick, P. W., & Hammond, A. L. (2005). Discovery of distal ejecta from the 1850 Ma Sudbury impact event: *Geology 33*, 193–196.
4. Litasov, K. D., & Ohtani, E. (2005). Phase relations in hydrous MORB at 18–26 GPa: Implications for heterogeneity of the lower mantle. *Phys Earth Planet Inter 150*, 239–263.
5. French, B. M., & Koeberl, C. (2010). The convincing identification of terrestrial meteorite impact structures: What works, what doesn't, and why. *Earth-Science* Reviews, *98*, 123–170.
6. Glikson, A. Y., Mory, A. J., Iasky, R. P., Pirajno. F., Golding, S. D, & Uysal, I., T. (2005). *Australian Journal of Earth Sciences, 52*, 545–553.
7. Kumar, R., Kumar, S., BadJena, S., & Mehrotra, S. P. (2005b). Hydration of mechanically activated granulated blast furnace slag. *Metallurgical and Materials Transactions B 36B*, 473–484.
8. Southworth, D. R., Barton, R. A., Verbridge, S. S., Ilic, B., Fefferman, A. D., Craighead, H. G., & Parpia, J. M. (Jun 2009). "Stress and Silicon Nitride: *A Crack in the Universal Dissipation of Glasses.*" *Physical Review Letters, 102*, 4.

CHAPTER 17

A RESEARCH NOTE ON OZONATION AS A METHOD OF UNSATURATED RUBBER MODIFICATION FOR IMPROVING ITS ADHESION PROPERTIES

V. F. KABLOV, N. A. KEIBAL, S. N. BONDARENKO,
D. A. PROVOTOROVA, and G. E. ZAIKOV

CONTENTS

ABSTRACT

This chapter considers the investigation related to modification of unsaturated rubbers by means of ozonation for improving their adhesion properties. Mechanism of macroradicals formation in the ozonation process as well as the influence of the contact time in the modification on adhesion properties of the rubbers has been studied. It has been shown that the best characteristics comparing with the initial ones can be obtained at 1 h ozonation. The gluing strength increases by 10–70% on average at that.

17.1 INTRODUCTION

Today, despite the existence of a large number of adhesives, which differ not only in the composition and properties, but also in manufacturing technology, formulations and intended purpose, the problem of creating new adhesives with a certain set of properties is still relevant. This is due to fact that glue compositions are imposed to ever more high demands related to operation conditions of construction materials and products.

The problem may be solved by applying the targeted modification of a film-forming polymer, which is a base component of any glue composition. Modification is of a priority than creating completely new adhesive formulations. The modification process is more advantageous from both an economic and technological points of view and allows not only to improve the performance of rubber, but also to maintain a basic set of properties.

As it is known, there are several methods of film-forming polymer modification. They are physical, chemical, photochemical modification, modification with biologically active systems and combinations of these methods.

Epoxydation, being one of the variants of chemical modification, represents a process of introduction of epoxy groups to a polymer structure that improves properties of this polymer. Materials based on epoxidized polymers show high physical and mechanical characteristics, meet the requirements to the strength and dielectric parameters, and manifest good adhesion to metals, which is achieved due to high adhesive activity of epoxy groups. Thanks to these properties they find application as coatings for metals and plastics, adhesives, mastics and potting compounds in electrotechnics, microelectronics and other areas of engineering [1].

Chlorinated natural rubber (CNR) is applied as an additive in glue compositions based on chloroprene rubber that, in its turn, is widely used in industry for gluing of different rubbers together with metals or each other [2]. As an individual brand glues based on CNR are rarely produced.

Isoprene rubber is an analog of natural rubber, used in the majority of rubber adhesives, but due to its low cohesive strength applied in their formulations much more rarely.

Thereby, the investigations concerned with the development of glue compositions based on these rubbers with improved adhesion characteristics to materials of different nature are of particular interest. It can be achieved by modification [3].

17.2   METHODICAL PART

In this chapter, a possibility of chlorinated natural rubber and isoprene rubber epoxydation by means of ozonation is investigated with the aim of improving adhesion characteristics of glues based on these rubbers as it is known that epoxy compounds are good film-formers in glue compositions and increase the overall viscosity of the last ones.

The introduction of epoxy groups in a polymer structure was carried out by means of ozonation as ozone is highly reactive towards double bonds, aromatic structures and C-H groups of a macrochain.

The contact time (0.5–2 h) was varied in the ozonation process. The rest parameters which are ozone concentration (5 $10^{-5\%}$ vol.) and temperature (23°C) were kept constant.

Further, glue compositions based on the ozonized rubbers were prepared.

Glue compositions based on the ozonized CNR were 20% solutions of the rubber in an organic solvent which was ethyl acetate. The compositions based on the isoprene rubber were 5% solutions of the ozonized rubber in petroleum solvent.

The gluing process was conducted at 18–25°C with a double-step deposition of glue and storage of the glued samples under a load of 2 kg for 24h. Glue bonding of vulcanizates was tested in 24 (±0,5) hours after constructing of joint by method called "Shear strength determination" (State Standard 14759-69), in quality of samples there were used poly isoprene (SKI-3), ethylenpropylen (SKEPT-40), butadiene-nitrile (SKN-18), and chloroprene (Neoprene) vulcanized rubbers.

17.3   RESULTS AND DISCUSSION

In ozonation, a partial double bonds breakage in the rubber macromolecules that leads to the macroradicals formation occurs (Fig. 17.1). Ozone molecules are attached to the point of the rubber double bonds breakage with the formation of epoxy groups [4]:

The formed macroradicals are probably interacting with macromolecules of the rubber which is a substrate material, thereby higher adhesion strength is provided.

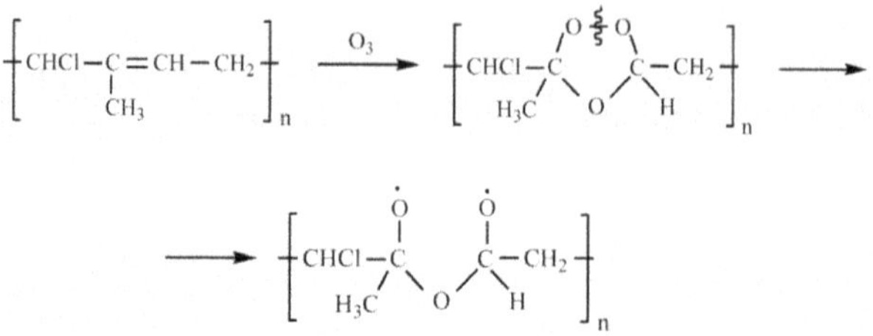

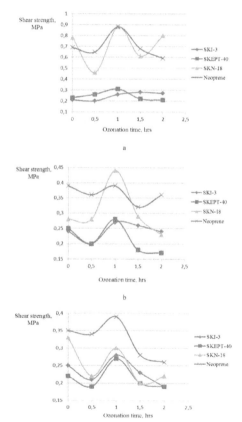

FIGURE 17.1   Reaction of the macroradicals formation (on the example of CNR).

At first, the CNR of three brands was investigated, they are as follows: S-20, CR-10 and CR-20. The results obtained in ozonation of three rubber brands are shown on the (Fig. 17.2).

FIGURE 17.2   Influence of ozonation time on shear strength at gluing of vulcanizates with glue compositions based on CNR of the S-20 (a), CR-10 (b) and CR-20 (c) brands accordingly.

A decrease in the adhesion strength at τ = 0.5 h may be connected with preliminary destruction of macromolecules under action of reactive ozone.

From the Fig. 17.2, we can see that the maximal figures correspond to 1hr ozonation. The adhesion strength for rubbers based on different caoutchoucs increases by 10–40% at that. The results of the shear strength change depending on the rubber brand and adhered type are shown on the (Fig. 17.3).

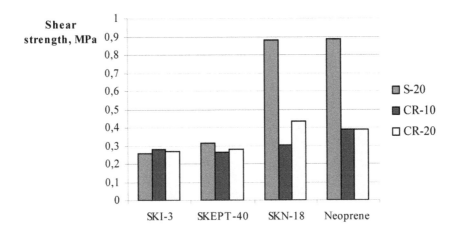

FIGURE 17.3   A change in shear strength for different CNR brands depending on adhered type (ozonation time τ = 1 h).

It should be noticed that the extreme nature of the above-mentioned dependences can be explained by the diffusion nature of the interaction between adhesive and substrate. As it's shown from the figures, with the increasing content of functional groups the strength began to reduce, having reached of certain limit in the adhesive. In this case, only adhesive molecules have ability to diffusion [5].

Isoprene rubber was also treated with ozone at the same parameters maintained for CNR. The results are on the Fig. 17.4.

The data in Fig. 17.4 confirm the ambiguity of the ozonation process. When the contact time is equal to 15 min (as in the case of CNR epoxydation [6]), possible preliminary destruction of the rubber macromolecules takes place, which on the graphs is proved by almost simultaneous reduction in the shear strength values. Concurrently, formation and subsequent growth of macroradicals go on, as evidenced by the increase in adhesive characteristics at ozonation time 0, 5 and 1hr. Here the shear strength at gluing of different vulcanized rubbers increases by 10–70% on average, and then it starts to reduce again.

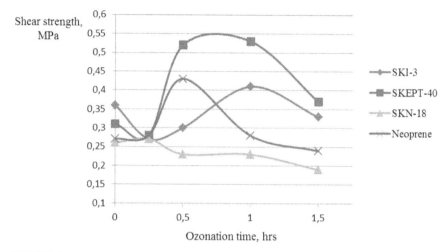

FIGURE 17.4   Influence of ozonation time on adhesive strength for the compositions based on isoprene rubber.

With further increase in ozonation time, the values of adhesion strength decline. That is apparently related to saturation of the polymer chain with epoxy groups and decrease in mobility of the macromolecules, and, consequently, the degree of interaction of the substrate with the adhesive composition as well as with destruction of the polymer chains.

17.4   CONCLUSIONS

Thus, ozonation can be applied as an effective method of enhancing adhesion properties of rubbers at the modification of film-forming polymers that are a main component in glues. Changing one of the parameters during the ozonation process we can obtain such content of epoxy groups at which the characteristics of adhesion strength will be maximal.

KEYWORDS

- **Glue compositions**
- **Gluing, vulcanizates**
- **Modification**
- **Ozonation**
- **Unsaturated rubbers**

REFERENCES

1. Solovyev, M. M. (2009). Local dynamics of oligobutadienes of different microstructure and their modification products: thesis of PhD in Chemistry Sciences: 02.00.06/ Solovyev Mikhail Mikhailovich. Yaroslavl, 201 p.
2. Dontsov, A. A., Lozovick, G. Ya., & Novitskaya, S. P. (1979). Chlorinated polymers Moscow: Khymiya, 232 p.
3. Kablov, V. F., Bondarenko, S. N., & Keibal, N. A. (2010). Modification of elastic glue compositions and coatings with element containing adhesion promoters: monograph. Volgograd: IUNL VSTU 238 p.
4. Zaikov, G. E. (2000). Why do polymers age. *Soros Educational Journal 6(12),* 52.
5. Berlin, A. A., & Basin, V. E. (1969). Basics of Polymer Adhesion Khymiya: Moscow, 320 p.
6. Keibal, N. A., Bondarenko, S. N., Kablov, V. F., Provotorova, D. A., & Gabriel, A., Ozonation of Chlorinated Natural Rubber and Studying its Adhesion Characteristics (2012). Rubber: Types, Properties and Uses. Popa, NY: Nova Publishers, 275–280.

CHAPTER 18

A RESEARCH NOTE ON THE INTUMESCENT FIRE AND HEAT RETARDANT COATINGS BASED ON PERCHLOROVINYL RESIN FOR FIBERGLASS PLASTICS

V. F. KABLOV, N. A. KEIBAL, S. N. BONDARENKO,
M. S. LOBANOVA, and A. N. GARASHCHENKO

CONTENTS

ABSTRACT

The purpose of research was obtaining fire retardant coatings based on perchlo-rovinyl resin with improved adhesive properties to protect fiberglass plastics. The article presents the results of studies on the influence of a modifier based on the phosphorus-boron-nitrogen-containing oligomer (PEDA) and filler, which is thermal expanded graphite on physical, mechanical and fire retardant properties of the coatings. It was found that the product PEDA is an effective fire retardant, and its introduction to the compositions provides fire resistance and high adhesion of the coatings. The dependence of the filler amount on fire retardant properties of the coatings, the ability to coking and coke strength were identified.

18.1   INTRODUCTION

In many cases polymer construction materials are a good alternative to metals and reinforced concrete. Still, it is known that the majority of such materials are combustible. That is why implementation of the materials to the building industry is associated with solving a range of engineering problems; one of them is providing the materials with the required fire safety. The fire hazard of polymers and composite materials is understood as a complex of properties, which along with combustibility includes the ability for ignition, lighting, flame spreading, quantitative evaluation of smoke generation ability, and toxicity of combustion products.

Fiberglass plastics find ever-widening applications in different industrial fields. The main benefit of fiberglass plastics is higher strength and lower density compared to metals; they are not subjected to corrosion.

However, together with the valuable property complex of fiberglass plastics, they also have a significant drawback that is a low resistance to open flame.

The sufficient increase in fire safety of fiberglass constructions may be achieved by using the passive protection measures – applying flame retardant in tumescent coatings.

Under the influence of high temperatures on a surface of the object protected from fire an in tumescent surface appears which obstructs penetration of heat and fire spread over the surface of the material.

For effective fire protection it is necessary to apply compounds which components inhibit combustion comprehensively: in a solid phase it is carried out by transforming the destruction process in a material, in a gaseous phase – preventing the oxidation of the degradation products [1, 2].

A standard formulation of a fire retardant coating includes an oligomer binder as also fire retarding nitrogen, phosphorus and/or halogen containing inorganic and organic compounds. The fire retarding effect is enhanced at the combination of different heteroatoms in an antipyrine [3, 4].

Previously, it was found that phosphorus boron containing compounds are effective antipyrenes in a fire retardant composition [5–7].

In the investigation we developed a new phosphorus-boron-nitrogen-containing oligomer (PEDA). The oligomer has a good compatibility to a polymer binder, slightly migrates from a polymer material, and is an effective antipyrine when has the lower phosphorus content [6, 7].

Phosphorus-boron-nitrogen-containing oligomer and polymer products comprising -P=O, -P-O-B-, -B-O-C- and -C-N-H- bonds are not studied enough. IR spectroscopy showed that these groups are a part of the PEDA macromolecule composition.

To improve physical and mechanical properties of coatings and characteristics of fire protection efficiency, the fire retardant coatings including the phosphorus-boron-nitrogen-containing oligomer PEDA as a modifier and based on perchlorovinyl resin (PVC resin) were obtained. The coatings are used for fiberglass plastics.

18.2   EXPERIMENTAL PART

With a purpose of defining the efficiency of the developed fire retardant coatings for fiberglass, a set of experiments was conducted.

The experiments on fire retardant properties were carried out on the developed technique by exposure of a coated fiberglass plastic sample to open flame. Time-temperature transformations on the nonheated surface of the fiberglass test sample was registered using a pyrometer measuring the moment of achieving the limit state a temperature of the fiberglass destruction beginning (280–300°C).

Then, the intumescence index of the coating was calculated. The intumescence index was determined by a relative increase in height of the porous coke layer compared to the initial coating height.

The coke residue was estimated by the relative decrease in the sample weight after keeping it in the electric muffle during 10, 20 and 30 min at 600°C.

For a possibility to use the fire retardant coatings, we need to solve a problem concerned with providing the required adhesion between the coating and protected material. The adhesive strength of the coatings to fiberglass plastic defined as the shear strength of a joint.

During the work, the studies on the combustibility and water absorption of the coatings were conducted.

The combustibility was evaluated by exposure of a sample to the burner flame (temperature peak 840°C) and fixing the burning and smoldering time after fire source elimination.

The experiments on the water absorption were performed in distilled water at temperature 23 ± 2°C for 24 h. The water absorption was estimated by a sample weight change before and after exposure to water.

The coke microstructure formed after a test sample burning was studied as well.

18.3   RESULTS AND DISCUSSION

As part of the research, the investigation of the coatings based on perchlorovinyl resin and containing the developed intumescent additive PEDA on fire protective properties was conducted. The results are presented in Table 18.1.

TABLE 18.1   Influence of PEDA Content on Fire Resistance of the Coatings Based on PVC Resin

Parameter	PEDA content, wt.%								
	Without coating	0		2.5		5.0		7.5	
Coating thickness, mm		0.7	1.0	0.7	1.0	0.7	1.0	0.7	1.0
Intumescence index	–	1.55	2.7	4.89	5.55	5.12	6.0	5.64	6.47
Time to the limit state, sec	18	29	32	44	52	48	57	55	63
Temperature of the non-heated sample side in 25 °C, sec	–	247	223	131	115	116	108	109	102

When a coating of 1 mm in thickness containing 7.5% PEDA (% of the initial composition weight) is used, the peak time to the limit state is established, and the intumescence index reaches 6.47 at that.

The temperature dependence of the nonheated sample side on flame exposure time at different PEDA content is shown in Fig. 18.1.

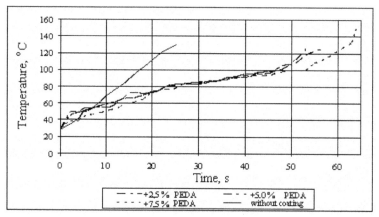

FIGURE 18.1   Dependence of temperature on the nonheated sample side on flame exposure time.

As it is seen in Fig. 18.1, the studied coatings allow keeping temperature on the nonheated sample side within the range 80–100°C for quite a long time; time to the limit state of test samples increased by 2–2.5 times.

Coke formation is an important process in fire and heat protection of a material. Achieving a higher intumescence ratio for carbonized mass, lower heat conductivity of coke and its sufficient strength, all these characteristics are the necessary conditions for effective fire protection.

The dependence of PEDA content influence on ability to form coke is presented on Fig. 18.2.

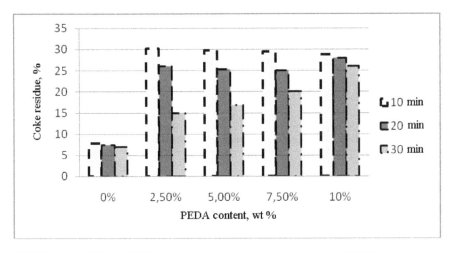

FIGURE 18.2   Effect of PEDA content on the coke residue values at 600°C.

As the diagram shows, with the growth of PEDA content the coke residue increases as well. It can be explained by the catalytic processes in coke formation caused by phosphorus boron containing substances [8].

In the experiments on combustibility it was found that the coatings containing PEDA are resistant to combustion and can be assigned the fire reaction class 1 as nonflammable (*see* Table 18.2).

TABLE 18.2   Influence of PEDA Content on Combustibility of the Coatings Based on PVC Resin

PEDA content, wt.%	Combustibility of a coating
0	Burning
2.5	Self-extinguishing in 2 sec
5.0	Self-extinguishing in a second
7.5	Not burning

The combustibility tests demonstrate that introducing PEDA into the compositions based on PVC resin promotes formation of a large coke layer; the coating film does not burn, because the presence of nitrogen in the modifying additive enables an enhancement of the fire and heat resisting effect.

TABLE 18.3   Influence of PEDA Content on Water Absorption of the Coatings Based on PVC Resin

PEDA content, wt.%	Extent of change in sample weight	pH
0	0.02	7
2.5	−0.05	5
5.0	−0.06	5
7.5	−0.05	5
10.0	−0.07	4

The results on determining the water absorption of the modified samples revealed an insignificant washout of PEDA that takes place through the slight diffusion of the modifying additive to the film surface that is evidenced by a change in pH in 24 h (Table 18.3). Nevertheless, this has no effect on fire resistance of the coatings.

As noted above, intumescent coatings should have good adhesion to the protected material; therefore, the studies on the influence of PEDA content on the adhesion strength of the coatings based on PVC resin to the fiberglass plastics were carried out while researching. The test results are illustrated by Fig. 18.3.

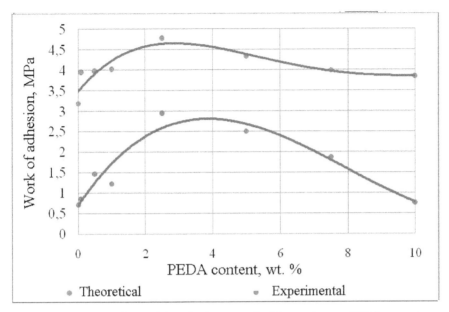

FIGURE 18.3 Dependence of the adhesion strength of the coatings on PEDA content.

So, it was established that the introduction of PEDA to the coating composition in amounts of 2.5–7.5% provides an increase in the adhesion strength by 1.5–4 times.

For confirmation of the experimental data, the work of adhesion was calculated according to the Young-Dupre equation. The surface tension was observed on the duNouy tensiometer. The contact angles of wetting were defined using a goniometric method.

The calculated values of the work of adhesion are well correlated with experimental data.

In order to improve intumescence and fire protection, the effect of introduced thermal expanded graphite (TEG), which served as filler, on coke formation and physical and mechanical characteristics of the coatings, was also studied in the chapter.

In a course of the study an optimal graphite amount was chosen so that the adhesion characteristics of the coatings would not become worse and allow obtaining a sufficiently hard coke.

The best results were achieved when applying PEDA and filler TEG. In this case the intumescence ratio reached 11.6 (see Fig. 18.4). The results on the influence of the coating modification and filler presence on the coke structure are presented in Figs. 18.5–18.7.

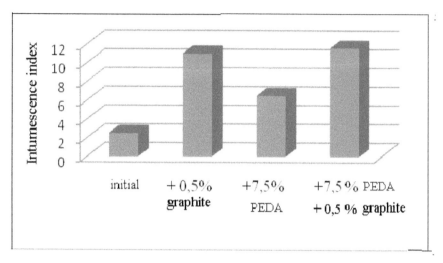

FIGURE 18.4   Influence of PEDA and TEG presence on the intumescence index of coatings.

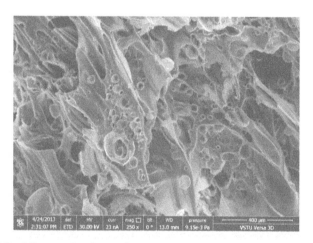

FIGURE 18.5   Micrograph of a coke structures on the initial coating at 250-fold magnification.

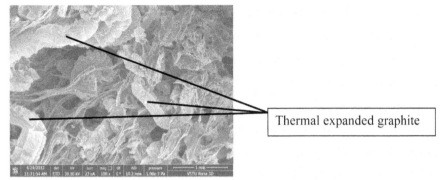

FIGURE 18.6   Micrograph of a coke structures on the coating containing TEG at 100-fold magnification.

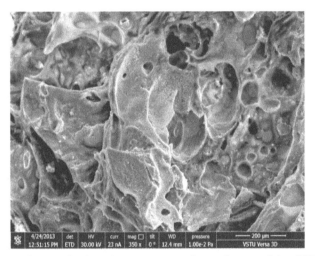

FIGURE 18.7   Micrograph of a coke structures on the coating containing PEDA and TEG at 350-fold magnification.

The foamed coke formed at the composition testing and not containing modifying additives and fillers has a coarse amorphous structure (Fig. 18.5); there are foamed globular formations of 10–100 μm in the coke volume grouping to associates.

In the compositions containing TEG only (Fig. 18.6), the coke structure is mainly determined by graphite that is present in the form of extended structures longer than 1000 μm, 50–100 times greater than the pore size of the foamed phase. The presence of these structures leads to increased friability of the foamed mass, coke has low strength.

In the coke structure of the coating containing PEDA and TEG (see Fig. 18.7) the extended structures, formed by graphite, disappear, and there are only short fragments of these formations. A consolidation of the carbon layers is observed which, probably, takes place due to formation of the high temperatures of polyphosphoric acids on the surface and between the layers of expanded graphite sites that solder layers, and, thereby, impede TEG intumescence; there is a slight shrinkage of the intumescent layer. As a result, the intumescence index of this composition is not substantially exceed the intumescence index of the composition containing only filler, but with a more ordered structure of coke and a sufficiently high strength and hardness of the composition the fire resistance increases. Such a coke can withstand more intense combustion gas streams.

18.4   CONCLUSION

Thus, the fire retardant coatings based on the developed phosphorus-boron-nitrogen-containing oligomer have high fire and heat protective and adhesive properties. The structure and presence of phosphorus, boron and nitrogen heteroatoms promotes an enhancement of the film-forming polymer carbonization and increase in the intumescence ration of the coatings. In addition, the definite advantage of PEDA application is that it is slightly washed out of a coating when exposed to water.

Introduction of the modifying additive PEDA in combination with a filler – thermal expanded graphite – permits to increase the coating intumescence by 11 times, resulting in improved fire and heat protective properties of the coatings and reduced destruction of fiberglass plastic.

The research has been done with financial support from Ministry of Education and Science of the Russian Federation under realization of the federal special-purpose program "Scientific, academic and teaching staff of innovative Russia" for 2009–2013 years: The Grant Agreement №14.B37.21.0837 "Development of active adhesive compositions based on element organic polymers and vinyl monomers."

KEYWORDS

- Adhesion
- Fiberglass plastics
- Filler
- Fire protection
- Fire retardant coatings
- Modifier
- Phosphorus-boron-nitrogen-containing oligomer

REFERENCES

1. Berlin, Al. Al. (1996). Combustion of Polymers and Polymer Materials of Reduced Combustibility. *Soros Educational Journal. 9*, 57–63.
2. *Shuklin, S. G., Kodolov, V. I., & Klimenko, E. N. (2004). Intumescent Coatings and the Processes that Take Place in Them Fibre Chemistry. 36(3). 200–205.*
3. Nenakhov, S. A. & Pimenova, V. P. (2011). Physical Transformations in Fire Retardant Intumescent Coatings Based on Organic and Inorganic Compounds ozharovzryvobezopasnost. *Fire and Explosion Safety. 20(8),* **17–24.**
4. Balakin, V. M., & Yu, E. **(2008).** Polishchuk. Nitrogen and Phosphorus Containing Antipyrenes for Wood and Wood Composite Materials Pozharovzryvobezopasnost *Fire and Explosion Safety. 17(2),* 43–51.
5. Shipovskiy, Ya. I., Bondarenko, S. N., & Yu, I. Goryainov. (2005). Fire Protective Modification of Wood. Proceedings of the International Scientific and Practical Conference. Dnepropetrovsk, *47,* 20.
6. Gonoshilov, D. G., Keibal, N. A., Bondarenko, S. N., & Kablov, V. F. (2010). Phosphorus Boron Containing Fire Retardant Compounds for Polyamide Fibers Proceedings of the 16th International Scientific and Practical Conference "Rubber industry: Raw materials. Manufactured materials. Technologies," Moscow, 160–162.
7. Lobanova, M. S., Kablov, V. F., Keibal, N. A., & Bondarenko, S. N. (2013). Development of Active Adhesive Fire and Heat Retardant Coatings for Fiber-Glass Plastics. All the Materials. *Encyclopaedic* Reference Book. *(04)* 55–58.
8. Korobeynichev, O. P., Shmakov, A. G., & Shvartsberg, V. M. (2007). *The Combustion Chemistry of Organophosphorus Compounds Uspekhi khimii. 76(11),* 1094–1121.

CHAPTER 19

A LECTURE NOTE ON INFLUENCE OF HIDROPHYLIC FILLER ON FIRE RESISTANCE OF EPOXY COMPOSITES

V. F. KABLOV, A. A. ZHIVAEV, N. A. KEIBAL, T. V. KREKALEVA, and A. G. STEPANOVA

CONTENTS

ABSTRACT

Fire resistant water containing epoxy composites have been developed and studied in the chapter. The influence of hydrophilic filler content on the fire resistance of epoxy polymers based on epoxy oligomers has been established.

19.1 INTRODUCTION

Today, expanding the range of technological and operational characteristics of composite materials based on epoxy oligomers is an urgent problem. Polymer composites based on epoxy resins are widely used as structural materials and adhesives. The advantages of epoxy composites are good adhesion to reinforcing elements, the lack of volatile by-products during hardening, and low shrinkage [1, 2].

However, in some cases, the use of epoxy composites is limited by their low thermal stability and fire resistance [3]. One of the benefits of epoxy resins is ability to regulate their composition by introducing various modifiers (fillers, plasticizers, fire retardants, etc.) resulting in materials with a given set of properties [4]. There are known fire retardant polymer compositions with microencapsulated fire extinguishing liquids (halogen phosphorus containing compounds, water, etc.). Microencapsulation substantially improves both technological and functional properties of the most diverse products and considerably expands the scope of their application.

19.2 EXPERIMENTAL PART

In the research the influence of hydrophilic filler content on the fire resistance of composites based on epoxy resin ED-20 was investigated.

Acrylamide copolymer POLYSWELL was applied as a hydrophilic filler. The acrylamide copolymer is a white granular material, density is 0.8–1.0 g/cm^3, water-swellable to form a polymer gel. In solutions the amide group shows weak-basic properties at the expense of the lone electron pair on the nitrogen atom that is the reason of nonchemical interaction of the polymer with water.

The compositions were obtained on the basis of epoxy resin by means of blending components as follows: epoxy resin ED-20, filler – acrylamide copolymer in the form of granules preliminary swollen in water in the ratio of 1:10, and hardener that was polyethylenepolyamine. The obtained reactive blends were molded and then hardened without heat supply for 24 h. The test samples had the following sizes: diameter is 50 mm, thickness is 5 mm.

For determining the efficiency of the developed composites, the experiments on the fire resistance by exposure of a sample to open flame using the universal Bunsen burner were conducted. With a help of the pyrometer C-300.3, measuring the moment of achieving the limit state, time-temperature transformations on the nonheated surface of the sample were registered.

Along with the fire resistance estimate of the developed compositions, the studies of samples on the water absorption and combustibility depending on the hydrophilic filler content were carried out.

The combustibility was evaluated by the standard technique on the rate of horizontal flame spread over the surface. A sample was exposed to the burner flame (temperature peak 840 °C) and the burning and smoldering time after fire source elimination was fixed.

The experiments on the water absorption were performed in distilled water at temperature 23±2 °C for 24 h. The water absorption was characterized by a sample weight change before and after exposure to water.

19.3   RESULTS AND DISCUSSION

As mentioned in the earlier section, the investigation on determining the fire resistant properties of epoxy compositions was carried out during the research. The results are shown in Figs. 19.1 and 19.2.

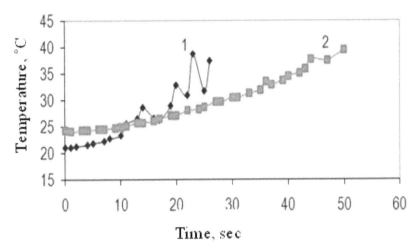

FIGURE 19.1   Dependence of temperature on the nonheated sample side on flame exposure time: 1 initial epoxy component; 2-epoxy composite containing hydrophilic filler.

As it is seen in Fig.19.1, the test sample damage takes place on the 15th sec, which is evidenced by temperature fluctuations on curve 1. The filled sample (contains 15% of the hydrophilic filler) maintains the integrity up to 50 sec; the sparking is observed at combustion that is, probably, related to water injection into the combustion zone. Besides, when the flame source is eliminated, the sample self-extinguishes for 2–3 sec.

The effect of hydrophilic filler content (5–20%) on the fire resistance of the composites was also studied in the chapter (Fig. 2).

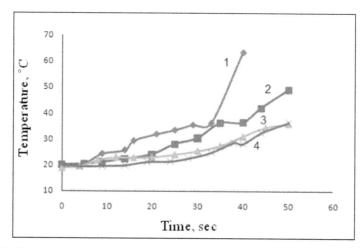

FIGURE 19.2   Dependence of temperature on the nonheated sample side on flame exposure time for epoxy composite containing hydrophilic filler in the following amounts: 5% (1), 10% (2), 15% (3) 20% (4).

When measuring temperature on the nonheated surface of water containing composites within a specified time span, it was established that the fire resistant properties improve with increasing hydrophilic filler content from 10% to 20%; the sample with filler content of 5% was damaged by 33 sec.

The results of experiments on the water absorption and fire resistance are presented in Table 19.1.

TABLE 19.1   Estimation of Water Absorption and Fire Resistance of Epoxy Composites

Parameter	Filler content in a composite, % by epoxy resin weight				
	Without filler	5	10	15	20
	Composition index				
Water absorption, %	0.41		0.33	0.24	0.28	0.28
Time to the limit state, sec	15.0		35.0	50.0	50.0	50.0
Temperature of the nonheated sample side in 25 sec, °C	Sample damaged		33.0	28.0	24.0	20.0

   Proposed compositions provide a significant increase in values of the fire resistance and water absorption compared to the initial sample. Time to the limit state for
the test samples goes up by 2.5 times at that.
   The data obtained in combustibility tests of water containing compositions are
illustrated in Fig. 19.3. The best results were obtained with filler content of 15 and
20%. In this way, the flame spread rate for the initial sample is 18 mm/min and,
when the hydrophilic filler is used, it is equal to 3 mm/min.

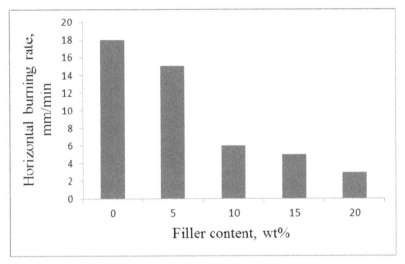

FIGURE 19.3   Evaluation of the horizontal flame spread rate over epoxy composite surface.

   On exposure to flame, there is a kind of microexplosions and injection of fire
extinguishing liquid (water) occurring in the combustion zone. In this case, combustion inhibition is likely due to the absorption of a significant heat amount characterized by the high heat capacity and high water evaporation heat. A possible factor in
reducing the flame-spread rate is also water displacement of the combustion reaction components from the reaction zone.

19.4   CONCLUSION

So, the properties of water containing epoxy composites were developed and investigated as well as the possibility of applying a hydrophilic filler as an additive that
increase the fire resistance of hardened epoxy compositions based on epoxy resin
ED-20 was shown in the chapter.

One of the important conclusions in this chapter is the high prospect of the modification method of epoxy composites in order to give them a set of specific properties using filled microencapsulated materials.

KEYWORDS

- **Acrylamide copolymer**
- **Epoxy composites**
- **Fire resistance**
- **Hydrophilic filler**

REFERENCES

1. Eselev, A. D., & Bobylev, V. A. (2006). State-of-the-art production of epoxy resins and adhesive hardeners in Russia. *Adhesives. Sealants, (7),* 2–8.
2. Amirova, L. M., Ganiev, M. M., & Amirov, R. R. (2002). *Composite Materials Based on Epoxy Oligomers.* Kazan: Novoe znanie plc, 167 p.
3. Kopylov, V. V., & Novikov, S. N. (1986). *Polymer Materials of Reduced Combustibility.* Moscow: Khimiya, 224 p.
4. Kerber, M. L., & Vinogradov, V. M. (2008). *Polymer Composite Materials: Structure, Properties, and Technology.* Ed. by Berlin, A. A. St. Petersburg: Professiya, 560 p.

CHAPTER 20

POLYELECTROLYTE MICROSENSORS AS A NEW TOOL FOR METABOLITES' DETECTION

L. I. KAZAKOVA, G. B. SUKHORUKOV, and L. I. SHABARCHINA

CONTENTS

ABSTRACT

Enzyme based micron sized sensing system with optical readout was fabricated by co-encapsulation of urease and dextran couple with pH sensitive dye SNARF-1 into polyelectrolyte multilayer capsules. Co-precipitation of calcium carbonate, urease and dextran followed up by multilayer film coating and Ca-extracting by EDTA resulted in formation of 3.5–4 micron capsules, what enable the calibrated fluorescence response to urea in concentration range from 10^{-6} to 10^{-1} M. Sensitivity to urea in concentration range of 10^{-5}–10^{-1} M was monitored on capsule assemblies (suspension) and on single capsule measurements. Urea presence can be monitored on single capsule level as illustrated by Confocal fluorescent microscopy.

20.1   INTRODUCTION

Design of micro and nanostructured systems for in-situ and in-vivo sensing become an interesting subject nowadays for biological and medical-oriented research [1–3]. Miniaturization of sensing elements opens a possibility for noninvasive detection and monitoring of various analyzes exploiting cell and tissues residing reporters. Typical design of such sensor is nano and micro particles loaded with sensing substances enable to report on presence of analyzes by optical means [4–9]. The particles containing fluorescent dye can be use as a sensor for relevant analyzes such as H^+, Na^+, K^+, and Cl^- et al. [10–13]. The fluorescent methods are most simple and handy among the possible ways of registration. They provide high sensitivity and relative simplicity of data read-out. For analysis of various metabolites it is necessary to use the enzymatic reactions to convert analyzes to optically detectable compound [14, 15]. In order to proper functioning all components of sensing elements (fluorescence dyes, peptides, enzymes) are to be immobilized in close proximity of each other. That "tailoring" of several components in one sensing entity represents a challenge in developing of a generic tool for sensor construct. One approach to circumvent problem has been introduced by the PEBBLE (Photonic Explorers for Bio-analyze with Biologically Localized Embedding) system [5, 16]. PEBBLE is a generic term to describe use co-immobilization of sensitive components in inert polymers, substantially poly acrylamide, by the micro emulsion polymerization technique [17]. This technique is useful for fluorescent probe, but to our mind, is too harsh for peptides and enzymes capsulation due to organic solvents involved in particle processing. Multilayer polyelectrolyte microcapsules have not this shortcoming as they are operated fully in aqueous solution at mild condition. These capsules are fabricated using the Layer-by-Layer (LbL) technique based on the alternating adsorption of oppositely charged polyelectrolyte onto sacrificial colloidal templates [18, 19]. Immobilization of one or more enzymes within polyelectrolyte microcapsules can be accomplished by the co precipitation of these enzymes into the calcium carbonate particles, followed by particle dissolution in mild condition leaving a set of protein retained in capsule [15, 20, 21]. A fluorescence dye can be included in polyelectrolyte capsules as well. Thus, the multilayer polyelectrolyte encapsulation technique microcapsules allows in principle combining enzyme activity for selected metabolite and registration ability of dyes in one capsule. In this chapter, urea detection is demonstrated using capsules containing Urease and pH sensitive dye.

The concentration of urea in biological solutions (blood, urine) is a major characteristic of the condition of a human organism. Its value may suggest a number of acute and chronic diseases: myocardial infarction, kidney and liver dysfunction. The measurement of urea concentration is a routine procedure in clinical practice. Urease based enzymatic methods are most widely used for urea detection and use Urease enzyme as a reactant. There are multiple Urease based methods, which differ from each other by the manner of monitoring the enzymatic reaction. Despite of high specificity, reproducibility and extremely sensitivity for urea as urea is the only physiological substrate for Urease, these methods are all laborious and time-consuming since freshly prepared chemical solutions and calibration are required daily. They all are lacking in-situ live monitoring what makes them inappropriate for analysis in vivo as residing sensors. Embedding of Urease into polyelectrolyte's microcapsules can help to solve these problems [22, 23]. The encapsulated Urease completely preserves its activity at least 5 days at the fridge storage [22].

Aim of this chapter was to demonstrate a particular example of a sensor system, which combines catalytic activity for urea and at the same time, enabling monitoring enzymatic reaction by optical recording. The proposed sensor system is based on multilayer polyelectrolyte microcapsules containing urease and a pH-sensitive fluorescent dye, which translates the enzymatic reaction into a fluorescently registered signal.

20.2 EXPERIMENTAL DETAILS

20.2.1 MATERIALS

Sodium poly(styrene sulfonate) (PSS, MW = 70–000) and poly(allylamine hydrochloride) (PAH, MW = 70–000), calcium chloride dehydrate, sodium carbonate, sodium chloride, ethylene diamine tetra acetic acid (EDTA), TRIS, maleic anhydride, sodium hydroxide (NaOH) and Bromocresol purple were purchased from Sigma-Aldrich (Munich, Germany). Urease (Jack bean, Canavalia ensiformis) was purchased from Fluka. SNARF-1 dextran (MW = 70,000) was obtained from Invitrogen GmbH (Molecular Probes #D3304, Karlsruhe, Germany). All chemicals were used as received. The bi-distillated water was used in all experiments.

20.2.2 PREPARATION OF SNARF-1 DEXTRAN AND SNARF-1 DEXTRAN/UREASE CONTAINING CACO₃ MICROPARTICLES

The preparation of loaded $CaCO_3$ microspheres was carried out according to the coprecipitation-method [20, 21]. To prepare the $CaCO_3$ microspheres loaded with SNARF-1 dextran were used: 1.6 mL H_2O, 0.5 mL 1 M $CaCl_2$, 0.5 mL 1 M Na_2CO_3 and 0.4 mL SNARF-1 dextran solution (1 mg/mL).

To prepare the $CaCO_3$ microspheres contained different ratio of SNARF-1 dextran and urease were used:

Sample I: 0.6 ml H_2O, 0.5 mL 1 M $CaCl_2$, 0.5 mL 1 M Na_2CO_3, 0.4 mL SNARF-1 dextran solution (1 mg/mL) and 1 ml urease (3 mg/mL);

Sample II: 0.8 mL H_2O, 0.5 mL 1 M $CaCl_2$, 0.5 mL 1 M Na_2CO_3, 0.2 mL SNARF-1 dextran solution (1 mg/mL) and 1 ml urease (3 mg/mL).

The solutions were rapidly mixed and thoroughly agitated on a magnetic stirrer for 30 s at 4 °C. After the agitation, the precipitate was separated from the supernatant by centrifugation (250 g, 30 s) and washed three times with water. The procedure resulted in highly spherical microparticles containing SNARF-1 dextran or SNARF-1 dextran and urease with an average diameter ranging from 3.5–4 μm.

20.2.3 FABRICATION OF SNARF-1 DEXTRAN LOADED MICROCAPSULES

Microcapsules were prepared by alternate layer-by-layer (LbL) deposition of oppositely charged polyelectrolyte poly (allylamine hydrochloride) (PAH, MW = 70 000) and poly (styrene sulfonate) (PSS, MW = 70 000) onto CaCO3 particles containing SNARF-1 dextran or SNARF-1 dextran and urease to give the following shell architecture: (PSS/PAH) 4 PSS. Short ultrasound pulses were applied to the sample prior to the addition of each polyelectrolyte in order to prevent particle aggregation. The decomposition of the CaCO3 core was achieved by treatment with EDTA (0.2 M, pH 7.0) followed by triple washing with water. The microcapsules were immediately subjected to further analysis or stored as suspension in water at 4 °C.

20.2.4 SPECTROSCOPIC STUDY

All spectroscopic studies were carried out with UV–vis spectrophotometer *Varian Cary 100* at constant agitation and thermostatic control at 20°C.

The SNARF-1 dextran concentrations in different capsules samples were estimated by matching absorption intensity of supernatant after coprecipitation of the dye with $CaCO_3$ to intensity of calibrated of SNARF-1 dextran concentrations in free solution. The average content of SNARF-1 dextran per capsule was calculated to be: (i) for SNARF-1 dextran $CaCO_3$ microparticales – 1 pg; (ii) for SNARF-1 dextran/urease $CaCO_3$ microparticles: *sample I*–0.6 pg, *sample II*–0.2 pg.

The amount of active urease immobilized into the polyelectrolyte microcapsules was determined under assumption that the enzyme retains its activity while encapsulated. Free urease had 100 U/mg according to the data sheets. The activities of free and encapsulated enzyme were determined from the decomposition of urea into two ammonia molecules and CO_2 using a pH-sensitive dye Bromocresol purple [24]. The urease aliquot solutions were added to a reaction mixture contained a necessary amount of urea and 0.015 mM Bromocresol, whose pH was apriory brought up to 6.2. The reaction kinetics was recorded as a change in the optical absorption of the dye at 588 nm to obtain the linear calibration plot. Then, the known number of microcapsules containing urease and SNARF-1 dextran was added to the reaction solution. The revealed activity of enzyme was compared with amount of free urease.

20.2.5   SPECTROFLUORIMETRIC STUDY

All spectrofluorimetric studies of SNARF-1 dextran and SNARF-1 dextran/urease were carried out with the spectrofluorimeter *Varian Cary Eclipse*, at constant agitation, thermostatic control at 20°C, λ_{exc}=540 nm, slit width: excitation at 10 nm and emission at 20 nm. The microcapsule suspensions were used at concentration $2 \cdot 10^6$ capsules/mL, which was estimated with the cytometer chamber. All solutions were prepared on bi-distilled water.

TRIS-maleate buffer solutions for pH setting were prepared by adding appropriate quantity of 0.2 NaOH to 0.2 M TRIS and maleic anhydride mixture and diluted to 0.05 M concentration.

20.2.6   CONFOCAL LASER SCANNING MICROSCOPY

Confocal images were obtained by Leica Confocal Laser Scanning Microscope TCS SP. For capsules visualization 100* oil immersion objective was used throughout. 10 μL of the SNARF–1-dextran/urease capsules suspension was placed on a cover slip. To this suspension 10 μL of 0.1-mol/L urea is added. After about 20 min confocal images were obtained. The red fluorescence emission was accumulated at 600–680 nm after excitation by the FITC-TRIC-TRANS laser at 543 nm.

20.3   RESULTS AND DISCUSSION

Degradation of urea ($CO (NH_2)_2$) is catalyzed by Urease and results in the shift of the medium pH into the alkaline range.

$$CO (NH_2)_2 + H_2O = CO_2 + 2 NH_3$$

Monitoring of the urea degradation can be done by using SNARF-1 as pH-sensitive dye to follow changes of the pH in the enzyme driven reaction. In order to fabricate the sensing microcapsule the urease and SNARF-1 bearing dextran were simultaneously co-precipitated to form $CaCO_3$ spherical particles 3.5–4 μm in size, containing both components urease and fluorescent dye SNARF-1 coupled dextran (MW = 70,000) [15]. Then the particles were coated by standard layer-by-layer protocol with nine alternating layers of oppositely charged polyelectrolytes PSS and PAH. The formed shell had a $(PSS/PAH)_4 PSS$ architecture. After the dissolution of $CaCO_3$ with the EDTA solution the obtained capsule samples contain an enzyme and a fluorescent dye in its cavity.

The dye and enzyme concentrations inside the polyelectrolyte capsule are predetermined essentially at the stage of formation of the $CaCO_3$/SNARF–1 dextran/urease conjugate micro particles. Obviously, the amount of both components of urease and SNARF-1 dextran in the capsules and their ration should play an important role while functioning of entire sensing system is concerned. However, the final composition of co-precipitated particles and later capsules in fabricated samples may be different, though the same initial concentration of components used while preparing co-precipitating particles. It depends on a number of factors: adsorption, capturing and distribution of the components among the $CaCO_3$ particles, the size and the number of the particles yielded [25]. These parameters might vary from one experiment to another and therefore, it makes problematic to obtain two capsule samples with exactly the same content of encapsulated substances while relying on single capsule detection. Yet, it is imperative to observe this condition to reproduce the efficiency of any sensor. Thus, we always run experiments with at least two samples of capsules in parallel produced independent and having the same parameters at preparation. One of major problem on single particle/capsule detecting is deviation of fluorescent intensity from one particle to another due to uneven fluorescent distribution over population of capsules. To avoid this bottleneck and to obtain a sensor whose reliability and efficiency would not depend on the concentration of the reacting and registering substances in single capsule we opted the SNARF-1 fluorescent dye for the present study (Fig. 20.1).

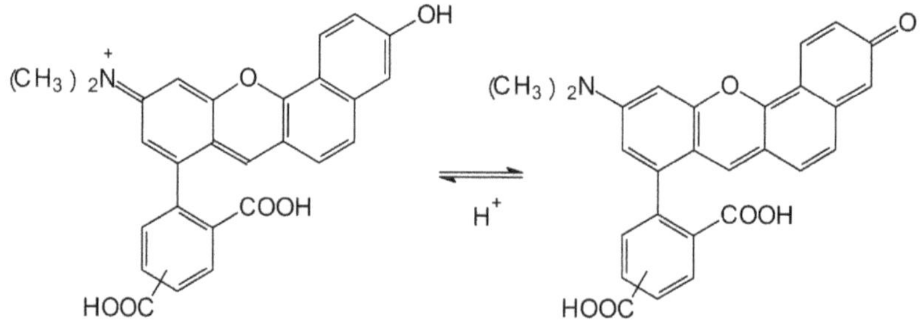

FIGURE 20.1   Structure of the protonated and deprotonated forms of the SNARF-1 dye.

The emission spectrum of SNARF-1 undergoes a pH-dependent wavelength shift from 580 nm in the acidic medium to 640 nm in alkaline environment. The ratio $R = I_{580 nm}/I_{640 nm}$ of the fluorescence intensities from the dye at two emission wavelengths allows to determine the pH value according to the ratio metric method. Dual emission wavelength monitoring is well established method eliminating a number of fluorescence measurement artifacts, including photo bleaching, sample's size thickness variation, measuring instrument stability and nonuniform loading of the indicator [26]. It becomes very important particularly if one is using not an ensemble of capsules but only single or few capsules in the analysis, for example, in the experiments with cells.

In order to verify a feasibility of fluorescence based urea sensing on two-component co-encapsulation we fabricated two polyelectrolyte capsule samples of the $(PSS/PAH)_4PSS$ shell architecture with different content of dye and urease. The first sample (*sample I*) contained in average 0.6 pg SNARF-1 dextran per capsule, while the content of the SNARF-1 dextran in the other sample was 0.2 pg per capsule (*sample II*). The concentration of active urease in samples was opposite 0.2 and 0.6 pg/capsule respectively what gives an average SNARF-1 dextran/urease ratio of 3:1 and 1:3 in these investigated samples.

Spectrofluoremetric studies were carried out to determine the correlation between the fluorescence intensity of the SNARF-1 dextran/urease capsules and the pH of the medium. Both the capsule samples were stored for 10 min in the 0.05 M TRIS-maleate buffer at pH in the range 5.5–9. The excitation wavelength was 540 nm. The capsules fluorescence spectra of the first sample are shown in Fig. 20.2. The spectra obtained for both samples were similar. The encapsulated dye is capable to provide information of the medium acidity in a reasonably wide range of pH. It is seen that fluorescence spectra are characteristic for every pH value. This fact can be used for calibration regardless amount of dye per capsules in studied samples.

The ratio between fluorescence intensity and pH can be described by the following equation according to Ref. [26]:

$$pH = pK_a - \log\left(\frac{R - R_{min}}{R_{max} - R} * \frac{I_{640nm}(B)}{I_{640nm}(A)} \right) \quad \text{(Equation 1)}$$

where $R = I_{580 nm}/I_{640 nm}$; R_{min} and R_{max} – are the minimal and maximal R values in the titration curve (Fig. 20.3, curves 3,4); $I_{640 nm}(A)$ and $I_{640 nm}(B)$-fluorescence intensities at 640 nm for the protonated and deprotonated forms of the dye, that is, in the acidic and alkaline media, respectively. The R_{min} and R_{max} meanings depend on the experimental conditions. In this study they were determined to be:

Sample 1: $R_{min} = 0.41$, $R_{max} = 1.96$, $I_{640 nm}(B)/I_{640 nm}(A) = 2$;
Sample 2: $R_{min} = 1.06$, $R_{max} = 2.14$, $I_{640 nm}(B)/I_{640 nm}(A) = 1.69$.

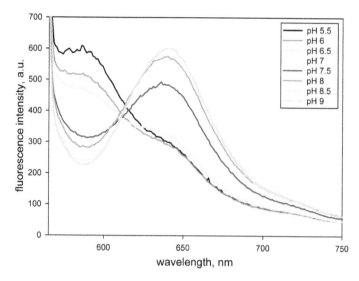

FIGURE 20.2 Fluorescence spectra of the SNARF-1 dextran/urease capsules in the 0.05 M TRIS-maleate buffer at pH in the range 5.5–9.

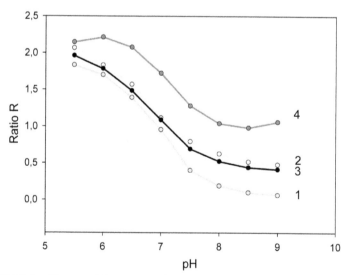

FIGURE 20.3 Change of fluorescence intensity ratio –R at 580 and 640 nm for: curve 1- SNARF-1 dextran water solution; curve 2 containing SNARF-1 dextran capsules; curve 3 SNARF-1 dextran/urease capsules (sample I, 0.6 pg dye/capsule; curve 4 SNARF-1 dextran/urease capsules (sample II, 0.2 pg dye/capsule) in 0.05 M TRIS-maleate buffer at pH in range 5.5–9.

To yield of the pK_a value the data were plotted as the log of the [H^+] versus the log $\{(R–R_{min})/(R_{max}–R)*(I_{640\ nm}B)/I_{640\ nm}A\}$. In this form, the data gave a linear plot with an intercept equal to the pK_a (Fig. 20.4, curves 3, 4). As follows from this data pK_a for sample 1 is equal 7.15, for the sample 2–7.25. Thus, the pK_a value differed on 0.1 for different samples whereas the concentration of dye in them differed in 3 times. However dependence of fluorescence intensity for the 2 sample was linear in the smaller interval of values.

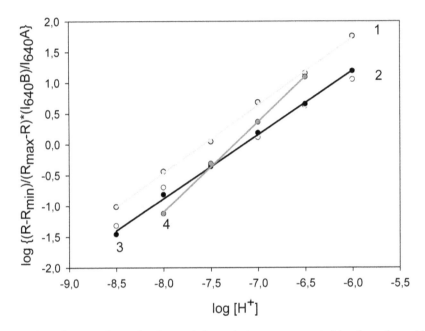

FIGURE 20.4 Experimental points and theoretical curves generated by the ratio metric method to determine the pK_a values: curve 1 for SNARF-1 dextran water solution, pK_a=7.58; curve 2 for containing SNARF-1 dextran capsules, pK_a=7.15; curve 3 for SNARF-1 dextran/urease capsules (sample I, 0.6 pg dye/capsule), pK_a=7.15; curve 4 for SNARF-1 dextran/urease capsules (sample II, 0.2 pg dye/capsule), pK_a=7.25.

In Figure 20.3, the curves 1, 2 shows the ratio R dependences on pH value for free fluorescent dye in comparison with SNARF-1 dextran capsules. The calculation of pK_a values (Fig. 20.4, the curves 1, 2) has demonstrated that for encapsulated dye it is less, than for the free dye solution. It is reasonable to assume that this effect is the result of the interaction between SNARF-1 dextran and polyelectrolyte shell, notably with PAA, because they have opposite charges.

Figure 20.5 illustrates the effect the urea in concentration from 10^{-6} to 10^{-2} mol/L produces on fluorescence spectra of capsules containing SNARF-1 dextran and urease. Urea concentration dependence is reflected spectroscopically by apparent pH change in the course of enzymatic reaction inside the capsules. The ammonium ions generated via enzymatic reaction in capsule interior effect on pH shift what is recorded by SNARF-1 on the plot (Fig. 20.5). The fluorescent spectra were measured at the 30-min time point after adding urea solutions at these concentrations to the SNARF-1 dextran/urease capsule's samples. Our particular attention was paid to the kinetics of the change at the fluorescence intensity ratio at 580 nm to 640 nm R (Fig. 20.6) and its relevance to amount of SNARF/urease. Parameter R was plotted versus time and as one can see on curves at high concentration of the urea substrate (10^{-3} M) level off at about 15–20 min after the beginning of the of the enzymatic reaction, while at low concentrations of urea (10^{-5} M) the time needed for flattening out spectral characteristics reaches 25–30 min. Remarkably, there are no substantial changes for samples at variable concentrations of the dye and Urease inside the capsule at least at studied range of 0.2–0.6pg per capsule. SNARF-1 dextran indicates only the course of the enzymatic reaction, therefore the time needed for the R parameter curve to level off correlates with the time needed to reach the equilibrium in the enzymatic reaction urea/urease occurring inside the capsule. Presumably, it takes about few minutes to equilibrate concentration of urea and its access to Urease. We assume rather fast diffusion of urea through the multilayer due to small molecular size of the molecules [27].

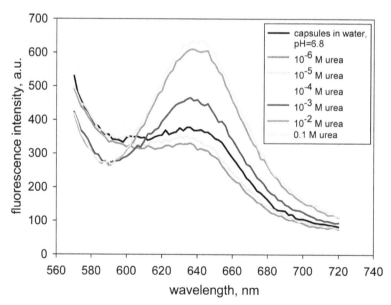

FIGURE 20.5 SNARF-1 dextran/urease capsule fluorescence spectra in water in the presence of urea from 10^{-6} to 0.1 M.

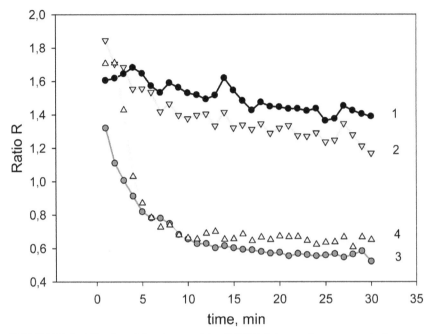

FIGURE 20.6   Kinetics of the change of the SNARF-1 dextran/urease capsule fluorescence intensity expressed through the fluorescence intensity ratio R at 580 and 640 nm in the presence of 10^{-5} M (curve 1 sample I, curve 2 sample II) and 10^{-3} M urea (curve 3 sample I, curve 4 sample II).

To calculate the apparent pH caused by urea concentration the values of R_{min}, R_{max}, pK_a and $I_{640\,nm}$ (B)/$I_{640\,nm}$ (A) were used. R_{max} was determined as a fluorescence intensity ratio at 580 nm and 640 nm in the capsules stored in bi-distilled water (pH = 6.4). R_{min} is the ratio of the fluorescence intensity spectrum related to the minimal R value at 580 nm and 640 nm in the SNARF-1 dextran/urease capsules when "high" concentration of urea (0.1 M) was added and 30 min after the enzymatic reaction begun. The pK_a value were assumed to be 7.15 and 7.25 for sample 1 and 2, respectively, that was in accord with the experiment with buffer solutions (Figs. 20.3 and 20.4). From the values obtained a calibration curve of the pH dependence inside the capsules on the urea concentration present in the solution was plotted. The calibration curve is presented in Fig. 20.7.

Thus, to determine the urea concentration in the solution it is necessary to obtain the following three spectra: (i) of the capsules without substrate (urea); (ii) of the capsules at "high" concentration of urea in the solution, for example, 0.1 M as used for our calibration plot; (iii) of the capsules in the investigated sample studied. These data will suffice to calculate the values of R_{min}, R_{max} and $I_{640\,nm}$ (B)/ $I_{640\,nm}$ (A),

which are characteristic of each particular sample of the sensing capsules. Using Eq.
(1), then can calculate the pH as apparently reached in the capsules in the course
of urea degradation and to compare its value with the calibration curve in Fig. 20.7

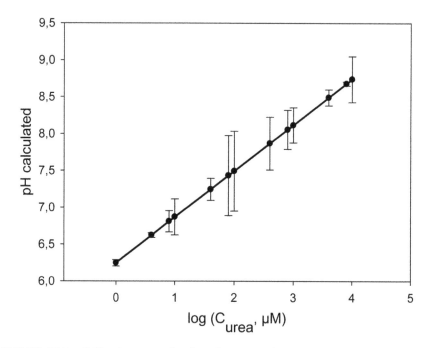

FIGURE 20.7   Calibration curve for detecting urea using the SNARF–1-dextran/urease
capsules in water solutions.

It is worth to notice, that this calibration curve is obtained for SNARF-1 dextran/
urease capsule in pure water without substantial contamination of any salt, which
could buffer the systems and spoil truly picture for urea detection. We carried out
experiments to build a similar calibration curve in the presence of the 0.001 M
TRIS-maleate buffer (were used solutions with the pH 6.5 and 7.5) but it resulted
in overwhelming effect of pH buffering. Buffering the solution eliminates the pH
change caused in a course of enzymatic reactions. Thus, it sets a limit for detection of urea concentration using SNARF-1-dextran/urease capsules. However the
calibration in conditions of particular experimental system is reasonable at salt free
solution assumption. Summarizing, one can state the presented in Fig. 20.7 calibration curve as suitable for estimation of urea concentrations in-situ in water solutions.

Feasibility studies on single capsule detection of urea presence were carried out
using Con focal Fluorescent Microscopy. CLSM image of SNARF-1 dextran/urease
capsules in absence and at 0.1 M urea added to the same capsules are presented on

Fig. 20.8. Small distinctions in the form and the sizes of capsule population are connected with nonuniformity of SNARF-1 dextran/urease CaCO$_3$ particles received in coprecipitation process what is rather often observed for calcium carbonate templated capsules containing proteins [20].

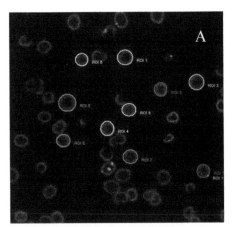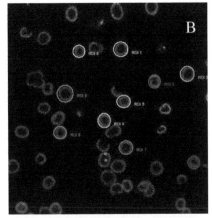

FIGURE 20.8 Confocal fluorescence microscopy images of (PSS/PAH)$_4$PSS capsules loaded with SNARF-1 dextran (MW= 70kDa) and urease enzyme in water (image A) and 0.1 M urea concentrations (image B). Table 20.1 shows the increase of Mean energy of individual capsules before (F$_{low}$) and after the addition 0.1 M urea solution (F$_{high}$) to water capsule suspension. The red fluorescence emission was accumulated at 600–680 nm after excitation by the FITC-TRIC-TRANS laser at 543 nm.

The images have been processed with Leica Confocal Laser Scanning Microscope TCS SP software to quantify the effect. The image areas corresponding to location of 10 selected capsules are set off in different colors -ROI$_{1-10}$ (Region of Interest). The value of average intensity of a luminescence is defined as parameter Mean Energy by formula:

$$I_{Mean}^2 = \frac{1}{N_{Pixel}} \sum_{Pixel} I_i^2$$

where I$_{mean}$ is the average image energy of ROI areas; N$_{pixel}$ is the total number of pixels that are included in the calculation; I$_i$ is the energy correspond for particular pixel.

The energy meanings of individual capsules are presented in the Table 20.1.

TABLE 20.1   The Change of Mean Fluorescence Intensity of 10 Selected Capsules in Presence of 0.1 M Urea

ROI	Mean Energy, I_{low}	Mean Energy, I_{high}	I_{high}/I_{low}
ROI 1	2947.95	6467.42	2.1939
ROI 2	4635.87	10316.34	2.2253
ROI 3	2424.13	5694.92	2.3493
ROI 4	4021.83	11161.52	2.7752
ROI 5	1666.76	3727.05	2.2361
ROI 6	3177.11	7237.68	2.2781
ROI 7	2237.61	5932.76	2.6514
ROI 8	3192.70	7158.54	2.2422
ROI 9	4100.76	8944.83	2.1813
ROI 10	2758.57	6127.32	2.2212

Although, value of fluorescence intensity is seen to be different for each capsule the more than double increase of integrated intensity is well pronounced in all monitored capsules upon addition of urea solution. The distribution of energy relation therefore remains almost constant for each capsule. These data on single capsules are in good agreement with data presented on (Fig. 20.5) obtained on entire capsule population. Indeed, integrated area under spectra for black (no urea) and light blue (0.1 M urea) with spectral range of 600–680 nm is about twice in difference. This fact demonstrates the principal applicability to use single capsule for carrying out analysis of urea presence.

20.4   CONCLUSION

In this study we have demonstrated a particular example of a sensor system, which combines catalytic activity for the substrate (urea) and at the same time enabling to monitor the enzymatic reaction by coencapsulated pH sensitive dye. Substrate sensitive enzyme urease was coencapsulated together with SNARF-1 coupled to dextran in multilayer microcapsules. Enzymatic activity was recorded by fluorescent changes caused by increasing of pH in course of enzymatic cleavage of urea as measured on population of capsules and on single capsule imaging by con focal fluorescent microscope. Suggested method can be used to measure the concentration of urea in solutions where the content of urea is fairly high (blood, urine) and also able to detect urea at concentration down to 10^{-5} M at non-buffering solution. Spectroscopic parameters of microencapsulated sensors were found stable in regardless ratio between urease and SNARF-dextran in concentration range of 0.2–0.6pg per capsules what encourages such as microencapsulated sensing system as robust. Although, that pH sensitivity of dye has a limitation to function in buffers the concept of coencapsulation of metabolite active enzymes and dyes sensitive to product of enzymatic reactions is illustrated to be workable in reasonable concentration range and applicable for single capsule based detecting.

The presented results prove the concept of feasibility of microencapsulated enzyme/dye systems for local metabolite sensing and optical online recording. These coencapsulating sensors have advantages over well known PEBBLE systems as they could sense the substances via extra enzymatic reaction what is much more prospective in term of analyzes to be monitored, especially in biological systems as cells and tissue. The micron size of the sensors will pave the way for producing and applicability of injectable and implantable sensing systems like 'a smart tattoo' or delivered to the cell or tissue and serving for on-line monitoring of various biological processes. Aspects of, considered here, urea sensing has a particular challenge to measure concentration of urea in vivo (in cells and tissue, for example, in skin epithelium), what remains subject of further research.

KEYWORDS

- **Biological processes**
- **Enzymatic reaction**
- **Microencapsulated enzyme/dye systems**
- **Nanostructured materials**
- **Polyelectrolyte multilayer capsules**
- **Sensor system**

REFERENCES

1. Francisco, J. (2009). Arregui, *Sensors Based on Nanostructured Materials*.
2. Fehr, M., Okumoto, S., Deuschle, K., Lager, I., Looger, L. L., Persson, J., Kozhukh, L., Lalonde, S., & Frommer, W. B. (2005). *Biochem Soc Trans.*, *33 (1)*, 287–290.
3. T., Griffin, G. D., Alarie, J. P., Cullum, B., Sumpter, B., & Noid, D. (2009). Summary *Nanomedicine*, *4(8)*, 967–979.
4. Sukhorukov, G. B., Rogach, A. L., Garstka, M., Springer, S., Parak, W. J., A. Muñoz-Javier, Kreft, O., Skirtach, A. G., Susha, A. S., Ramaye, Y., Palankar, R., & Winterhalter, M. *Small, 3(6)*, 944–955.
5. Lee, Y. E., Smith, R., & Kopelman, R. (2009). *Annu. Rev. Anal Chem (Palo Alto Calif)*, *2*, 57–76.

6. Sukhorukov, G. B., Rogach, A. L., Zebli, B., Liedl, T., Skirtach, A. G., Köhler, K., Antipov, A. A., Gaponik, N., Susha, A. S., Winterhalter, M., & Parak, W. J. (2005). *Small*, *1(2)*, 194–200.

7. De Geest, B. G., De Koker, S., Sukhorukov, G. B., Kreft, O., Parak, W. J., Skirtach, A. G., Demeester, J., De Smedt, S. C., & Hennink, W. E. (2009). *Soft Matter, 5*, 282.

8. J. Peteiro-Cartelle, M., Rodríguez-Pedreira, Zhang, F., Rivera Gil, P., del Mercato, L. L., & Parak, W. J. (2009). *Nanomedicine, 4 (8)*, 967–979.

9. Sailor, M. J., & Wu, E. C. (2009). *Advanced Functional Materials, 19 (20)*, 3195–3208.

10. Nayak, S., & McShane, M. J. (2006). *Sensor Letters*, *4*, 433–439.

11. Kreft, O., Muñoz Javier, A., Sukhorukov, G. B., & Parak, W. J. *J. Mater. Chem. (42)*, 4471–4476.

12. Brown, J. Q., & McShane, M. J. (2005). *IEEE Sensors Journal*, *5*, 1197–1205.

13. del Mercato, L. L., Abbasi, A. Z., & Parak, W. J. (2011). *Small*, DOI: 10.1002/smll.201001144

14. Brown, J. Q., & McShane, M. J. (2005). *Biosensors and Bioelectronics, 21*, 1760–1769.

15. Stein, E. W., Volodkin, D. V., McShane, M. J., & Sukhorukov, G. B. (2006). *Biomacromolecules*, *7*, 710–719.

16. Brasuel, M., Aylott, J. W., Clark, H., Xu, H., R., Kopelman Hoyer, M., Miller, T. J., Tjalkens, R., & Philbert, M. (2002). *Sensors and Materials, 14*, 309–338.

17. Xu, H., Aylott, J. W., & Kopelman, R. (2002). *Analyst*, *127*, 1471–1477.

18. Donath, E., Sukhorukov, G. B., Caruso, F., Davis, S. A., & Möhwald, H. (1998). *Angew. Chem.*, Int. Ed., *37*, 2202–2205.

19. Sukhorukov, G. B., Donath, E., Davis, S., Lichtenfeld, H., Caruso, F., Popov, V. I., & Möhwald, H. (1998). *Polym. Adv. Technol. 9*, 759–767.

20. Petrov, A. I., Volodkin, D. V., & Sukhorukov, G. B. (2005). *Biotechnol Prog 21*, 918–925.

21. Sukhorukov, G. B., Volodkin, D. V., Gunther, A. M., Petrov, A. I., Shenoy, D. B., & Möhwald, H. (2004). *Journal of Materials Chemistry, 14*, 2073–2081.

22. Lvov, Y., Antipov, A. A., Mamedov, A., Möhwald, H., & Sukhorukov, G. B. (2001). *Nano Lett 1*, 125.

23. Lvov, Y., & Caruso, F. (2001). *Anal. Chem. 73*, 4212.

24. Paddeu, S., Fanigliulo, A., Lanzin, M., Dubrovsky, T., & Nicolini, C. (1995). *Sens. Actuators, 25*, 876–882.

25. Halozana, D., Riebentanz, U., Brumen, M., & Donath, E. (2009). Colloids and Surfaces A: *Physicochem. Eng.* Aspects, *342*, 115–121.

26. Whitaker, J. E., Haugland, R. P., & Prendergast, F. G. (1991*). Anal Biochem, 194 (2)*, 330–44.

27. Antipov, A. A., Sukhorukov, G. B., Leporatti, S., Radtchenko, I. L., Donath, E., & Möhwald, H. (2002). *Colloids and Surfaces A: Physicochemical and Engineering Aspects, 198–200*, 535–541.

CHAPTER 21

SIMULATION IN CLASSICAL NANOMATERIALS: NEW DEVELOPMENT AND ACHIEVEMENTS

SHIMA MAGHSOODLOU and AREZOO AFZALI

CONTENTS

ABSTRACT

In this study the principles of simulation for nanoscience are reviewed, in detail. For this purpose, the survey is started by introducing system. Systems consist of interacting, interrelated, or interdependent elements groups. For studying a system a model must be built, first. A model is a representation of an object, a system, or an idea in some form other than that of the entity itself. A model must be had a proper balance between accuracy and complexity. Simulation is a numerical technique for conducting experiments on a digital computer, which involves certain types of mathematical and logical models that describes the behavior of a business or economic system (or some component) over extended periods of real time. At the last part of this study, kinetic is chosen as a situation for modeling due to get an applicable view to use all necessary concepts of simulation.

21.1 INTRODUCTION

A series of advances in a variety of complementary areas cause a real progress in nanotechnology, for instance: the discoveries of atomically precise materials such as nanotubes and fullerenes, the ability of the scanning probe and the development of manipulation techniques to image and manipulate atomic and molecular configurations in real materials, the conceptualization and demonstration of individual electronic and logic devices with atomic or molecular level materials, the advances in the self-assembly of materials to be able to put together larger functional or integrated systems, and finally the advances in computational nanotechnology, physics and chemistry based modeling and simulation of possible nanomaterials, devices and applications. It turns out that at the nanoscale, devices and systems sizes have condensed sufficiently small, so that, it is possible to describe their behavior justly identically. The simulation technologies have become also predictive in nature, and many pioneer concepts and designs, which have been first proposed, based on modeling and simulations, and then are followed by their realization or verification through experiments. Microscopic analysis methods are needful in order to get new functional materials and study physical phenomena on a molecular level in the nanotechnology science. These methods treat the constituent species of a system, such as molecules and fine particles. Macroscopic and microscopic quantities of interest are derived from analyzing the behavior of these species [1–3]. These approaches, called "molecular simulation methods," are represented by the Monte Carlo (MC) and molecular dynamics (MD) methods. MC methods exhibit a powerful ability to analyze thermodynamic equilibrium, but are unsuitable for investigating dynamic phenomena. MD methods are useful for thermodynamic equilibrium but are more advantageous for investigating the dynamic properties of a system in a nonequilibrium situation [4, 5].

21.2   MODELING AND SIMULATION PRINCIPLES

21.2.1   SYSTEMS

21.2.1.1   WHAT IS A SYSTEM?

A system is a set of components, which are related by some form of interaction which act together to achieve some objective or purpose. In a system, components are the individual parts or elements that collectively make up the system and relationships are the cause-effect dependencies between components. An objective is the desired state or outcome, which the system is attempting to achieve [6].

21.21.1.2   STUDY A SYSTEM

A detailed study to determine whether, to what extent, and how automatic data-processing equipment should be used. It usually includes an analysis of the existing system and the design of the new system, including the development of system specification which provides a basis for the selection of the equipment (Fig. 21.1). There are some common steps for studying the behavior of a system [7]:
1. observe the behavior of a system;
2. formulate a hypothesis about system behavior;
3. design and carry out experiments to prove or disprove the validity of the hypothesis;
4. often a model of the system is used;
5. measure/estimate performance;
6. improve operation;
7. prepare for failures.

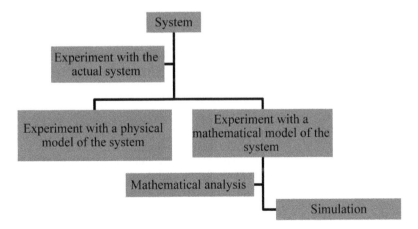

FIGURE 21.1   A schematic of a system.

21.2.1.3   AN EXAMPLE OF THE SYSTEM COMPONENTS DETERMINATION

Sets of interacting components or entities operating are collected together to achieve a common goal or objective. For example, in a manufacturing system its components are the machine centers, inventories, conveyor belts, production schedule and items produced and a telecommunication system is made of the messages, communication network servers [8, 9].

21.2.1.4   VARIOUS TYPES OF SYSTEMS

Systems can be classified in a variety of ways. There are natural and artificial systems, adaptive and nonadaptive systems. An adaptive system reacts to changes in its environment, whereas a nonadaptive system does not. Analysis of an adaptive system requires a description of how the environment induces a change of state. Now, some various types of system are reviewed in the following subsections.

21.2.1.4.1   NATURAL VS. ARTIFICIAL SYSTEMS

A natural system exists as a result of processes occurring in the natural world (e.g., river, universe) and an artificial system owes its origin to human activity (e.g., space shuttle, automobile) [10, 11].

21.2.1.4.2   STATIC VS. DYNAMIC SYSTEMS

A static system has structure but no associated activity (e.g., bridge, building) and a dynamic system involves time-varying behavior for complex systems (e.g., machine, U.S. economy). It deals with internal feedback loops and time delays that affect the behavior of the entire system [12, 13].

21.2.1.4.3   OPEN-LOOP VS. CLOSED-LOOP SYSTEMS

In all systems there will be an input and an output. Inputs are variables that influence the behavior of the system and outputs are variables, which determined by the system and may influence the surrounding environment. Signals flow from the input through the system and product an output.

An open-loop system cannot control or adjust its own performance but a closed-loop system controls and adjusts its own performance in response to outputs generated by the system through feedback. Feedback is the system function that obtains data on system performance (outputs), compares the actual performance to the desired performance (a standard or criterion), and determines the corrective action necessary. The controller acts on the error signal and uses the information to product the signal that actually affects the system, which we are trying to control [14] (Fig. 21.2).

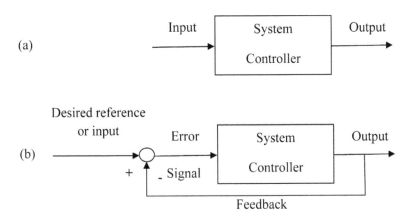

FIGURE 21.2   Graphical representations of the open loop (a) and close loop (b) systems.

We consider both internal and external relationships. The internal relationships connect the elements within the system, while the external relationships connect the elements with the environment, that is, with the world outside the system [15, 16].

The system is influenced by the environment through the input it receives from the environment. When a system has the capability of reacting to changes in its own state, we say that the system contains feedback. A nonfeedback, or open-loop, system lacks this characteristic.

The attributes of the system elements define its state. If the behavior of the elements cannot be predicted exactly, it is useful to take random observations from the probability distributions and to average the performance of the objective. We say that a system is in equilibrium or in the steady state if the probability of being in some state does not vary in time. There are still actions in the system, that is, the system can still move from one state to another, but the probabilities of its moving from one state to another are fixed. These fixed probabilities are limiting probabilities that are realized after a long period of time, and they are independent of the state in which the system started. A system is called stable if it returns to the steady state after an external shock in the system. If the system is not in the steady state, it is in a transient state [15].

21.3 MODELS

The first step in studying a system is building a model. A model is a representation of an object, a system, or an idea in some form other than that of the entity itself. A critical step in building the model is constructing the objective function, which is a mathematical function of the decision variables. For the complex systems a model describes the behavior of systems by using the construct theories or hypotheses which could be accounted for the observed behavior. So the model can predict future behavior and the effects that will be produced by changes in the system due to the analysis of the proposed systems.

21.3.1 MODELING APPROACH

Computational models, which also called simulation models, used to design new systems study and improve the behavior of existing systems. They allow the use of an interactive design methodology (sometimes called computational steering) so used in most branches of science and engineering (Fig. 21.3).

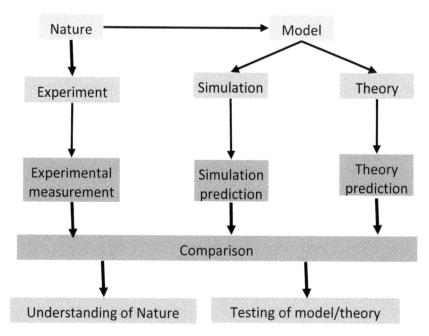

FIGURE 21.3 Modeling and experimental.
    There are many types of models, such as,

1. A scale model of the real system, for example, a model aircraft in a wind tunnel or a model railway [17].
2. A physical model in different physical system to the real one, for example, colored water in tubes has been used to simulate the flow of coal in a mine. More common in the use of electrical circuits-analog computers are based on this idea [18].
3. Mathematical Model: A description of a system where the relationship between variables of the system is expressed in a mathematical form [19, 20].
4. Deterministic vs. stochastic models: In deterministic models, the input and output variables are not subject to random fluctuations, so that the system is at any time entirely defined by the initial conditions and in stochastic models, at least one of the input or output variables is probabilistic or involves randomness [21, 22].

Among the models mathematical models are the most applicable models which have many advantages such as [23]:

1. enable the investigators to organize their theoretical beliefs and empirical observations about a system and to deduce the logical implications of this organization;
2. lead to improved system understanding;
3. bring into perspective the need for detail and relevance;
4. expedite the analysis;
5. provide a framework for testing the desirability of system modifications;
6. allow for easier manipulation than the system itself permits;
7. Permit control over more sources of variation than a direct study;
8. An additional advantage is that a mathematical model describes a problem more concisely than a verbal description does.

To build a model, important factors that act on the system must be included and unimportant factors that only make the model harder to build, understand, and solve should be omitted. For a continuous model, a set of equations that describe the behavior of a system as a continuous function of time t are written. Models use statistical approximations for systems that cannot be modeled using precise mathematical equations. While building a model it must be taken to ensure that it remains a valid representation of the problem. In order to get this purpose, a scientific model necessarily embodies elements of two conflicting attributes-realism and simplicity [24].

21.3.2 COMPUTATIONAL MODELS, ACCURACY, AND ERRORS

Appropriate balance between accuracy and complexity must be obtained in a model. An accurate representation of the physical system must be simple enough to implement as a program and solve on a computer in a reasonable amount of time. On the one hand, the model should serve as a reasonably close approximation to the real system and incorporate most of the important aspects of the system. On the other hand, the model must not be so complex that it is impossible to understand and manipulate. Adding details to the model makes the solution more difficult and converts the method for solving a problem from an analytical to an approximate numerical one [25].

In addition, it is not obligatory for the model to approximate the system to demonstrate the measure of effectiveness for all various alternatives. It needs to be a high correlation between the prediction by the model and what would actually happen with the real system. To specify whether this requirement is satisfied or not, it is important to test and establish control over the solution.

Usually, the model by reexamining the formulation of the problem and revealing possible flaws must be tested. Another touchstone for judging the validity of the model is determining whether all mathematical expressions are dimensionally consistent. A third useful test consists of varying input parameters and checking that the output from the model behaves in a plausible manner. The fourth test is the so-called retrospective test. It involves using historical data to reconstruct the past and then determining how well the resulting solution would have performed if it had been used. Comparing the effectiveness of this hypothetical performance with what actually happened then indicates how well the model predicts the reality. However, a disadvantage of retrospective testing is that it uses the same data that guided formulation of the model. Unless the past is a true replica of the future, it is better not to resort to this test at all.

Suppose that the conditions under which the model was built change. In this case the model must be modified and control over the solution must be established. Often, it is desirable to identify the critical input parameters of the model, that is, those parameters subject to changes that would affect the solution, and to establish systematic procedures to control them. This can be done by sensitivity analysis, in which the respective parameters are varied over their ranges to determine the degree of variation in the solution of the model [25].

After constructing a mathematical model for the problem under consideration, the next step is to derive a solution from this model. There are analytic and numerical solution methods. An analytic solution is usually obtained directly from its mathematical representation in the form of formula.

A numerical solution is generally an approximate solution obtained as a result of substitution of numerical values for the variables and parameters of the model. Many numerical methods are iterative, that is, each successive step in the solution uses the results from the previous step [26].

21.4 SIMULATION

The process of conducting experiments on a model of a system in lieu of either (i) direct experimentation with the system itself, or (ii) direct analytical solution of some problem associated with the system. A simulation of a system is the operation of a model of the system, as an imitation of the real system. A tool to evaluate the performance of a system, existing or proposed, under different configurations of interest and over a long period of time so a simulation of an industrial process is to learn about its behavior under different operating conditions in order to improve the process. Simulation is indeed an invaluable and very versatile tool in those problems where analytic techniques are inadequate. Simulation is a numerical technique for conducting experiments on a digital computer, which involves certain types of mathematical and logical models that describes the behavior of a business or economic system (or some component thereof) over extended periods of real time. Simulation does not require that a model be presented in a particular format. It permits a considerable degree of freedom so that a model can bear a close correspondence to the system being studied. The results obtained from simulation are much the same as observations or measurements that might have been made on the system itself [7, 27].

21.4.1 REASONS FOR SIMULATION

Simulation allows experimentation, although computer simulation requires long programs of some complexity and is time consuming. Yet what are the other options? The answer is direct experimentation or a mathematical model. Direct experimentation is costly and time consuming, yet computer simulation can be replicated taking into account the safety and legality issues. On the other hand, one can use mathematical models yet mathematical models cannot cope with dynamic effects. Also, computer simulation can sample from nonstandard probability distribution. One can summarize the advantages of computer simulation [28, 29].

Simulation, first, allows the user to experiment with different scenarios and, therefore, helps the modeler to build and make the correct choices without worrying about the cost of experimentations. The second reason for using simulation is the time control. The modeler or researcher can expand and compress time, just like pressing a fast forward button on life. The third reason is like the rewind button: seeing a scene over and over will definitely shed light on the answer of the question, "why did this happen?"

The fourth reason is "exploring the possibilities." Considering that the package user would be able to witness the consequences of his/her actions on a computer monitor and, as such, avoid jeopardizing the cost of using the real system; therefore, the user will be able to take risks when trying new things and diving in the decision pool with no fears hanging over her/his neck.

As in chess, the winner is the one who can visualize more moves and scenarios before the opponent does. In business the same idea holds. Making decisions on impulse can be very dangerous, yet if the idea is envisaged on a computer monitor then no harm is really done, and the problem is diagnosed before it even happens. Diagnosing problems is the fifth reason why people need to simulate [30].

Likewise, the sixth reason tackles the same aspect of identifying constraints and predicting obstacles that may arise, and is considered as one major factor why businesses buy simulation software. The seventh reason addresses the fact that many times decisions are made based on "someone's thought" rather than what is really happening.

When studying some simulation packages, the model can be viewed in 3-D. This animation allows the user "to detect design flaws within systems that appear credible when seen on paper or in a 2-D CAD drawing." The ninth incentive for simulation is to "visualize the plan."

It is much easier and more cost effective to make a decision based on predictable and distinguished facts. Yet, it is a known fact that such luxury is scarce in the business world [28].

Nevertheless, before trying out the "what if" scenario many would rather have the safety net beneath them. Therefore, simulation is used for "preparing for change" [29].

In addition, the 13th reason is evidently trying different scenarios on a simulated environment; proving to be less expensive, as well as less disturbing, than trying the idea in real life. Therefore, simulation software does save money and effort, which denotes a wise investment. In any field listing the requirements can be of tremendous effort, for the simple reason that there are so many of them. As such, the 14th reason crystallizes in avoiding overlooked requirements and imagining the whole scene, or the trouble of having to carry a notepad to write on it when remembering a forgotten requirement. While these recited advantages are of great significance, yet many disadvantages still show their effect, which are also summarized. The first hardship faced in the simulation industry is [28, 29, 31].

21.4.2   DANGERS OF SIMULATION

Becoming too enthusiastic about a model and forget about the experimental frame. Force reality into the constraints of a model and forget the model's level of accuracy. Also it should not be forgotten that all models have simplifying assumptions. If two modelers work together and cannot agree on a model, which can be due to the human nature, a consequence of it is the difficulty of interpreting the results of the simulation. It is the simple fact that simulation is not the solution for all problems. Hence, certain types of problems can be solved using mathematical models and equations.

Simulation may be used inappropriately Simulation is used in some cases when an analytical solution is possible, or even preferable. This is particularly true in the case of small queuing systems and some probabilistic inventory systems, for which closed form models (equations) are available.

Although of all dangerous of simulation, recent advances in simulation methodologies, availability of software, and technical developments have made simulation one of the most widely used and accepted tools in system analysis and operations research [32].

21.4.3   MODEL TRAINING

Simulation models can provide excellent training when designed for that purpose. Used in this manner, the team provides decision inputs to the simulation model as it progresses. The team, and individual members of the team, can learn from their mistakes, and learn to operate better. Moreover, training any team using a simulated environment is less expensive than real life. Some of model building requires special training. Model building is an art that is learned over time and through experience. Furthermore, if two models of the same system are constructed by two competent individuals, they may have similarities, but it is highly unlikely that they will be identical [33].

21.4.4   SIMULATION APPROACHES

There are four significant simulation approaches or methods used by the simulation community:
1.   Process interaction approach.
2.   Event scheduling approach.
3.   Activity scanning approach.
4.   Three-phase approach.

There are other simulation methods, such as transactional-flow approach, that are known among the simulation packages and used by simulation packages like Pro Model, Arena, Extend, and Witness. Another, known method used specially with continuous models is stock and flow method [34–39].

21.4.5   PROCESS INTERACTION APPROACH

The simulation structure that has the greatest intuitive appeal is the process interaction method. In this method, the computer program emulates the flow of an object (for example, a load) through the system. The load moves as far as possible in the system until it is delayed, enters an activity, or exits from the system. When the load's movement is halted, the clock advances to the time of the next movement of any load.

This flow, or movement, describes in sequence all of the states that the object can attain in the system. In a model of a self-service laundry, for example, a customer may enter the system, wait for a washing machine to become available, wash his or her clothes in the washing machine, wait for a basket to become available, unload the washing machine, transport the clothes in the basket to a dryer, wait for a dryer to become available, unload the clothes into a dryer, dry the clothes, unload the dryer, and then leave the laundry. Each state and event is simulated. Process interaction approach is used by many commercial packages [40].

21.4.6   TRANSACTION FLOW APPROACH

Transaction flow approach was first introduced by GPSS in 1962 as stated by Henriksen (1997). Transaction flow is a simpler version of process interaction approach, as the following clearly states: world-view was a cleverly disguised form of process interaction that put the process interaction approach within the grasp of ordinary users." In transaction flow approach models consist of entities (units of traffic), resources (elements that service entities), and control elements (elements that determine the states of the entities and resources). Discrete simulators, which are generally, designed for simulating detailed processes, such as call centers, factory operations, and shipping facilities, rely on such approach [40–42].

21.4.7   EVENT SCHEDULING APPROACH

The basic concept of the event scheduling method is to advance time to the moment when something happens next (that is, when one event ends, time is advanced to the time of the next scheduled event). An event usually releases a resource. The event then reallocates available objects or entities by scheduling activities, in which they can now participate. For example, in the self-service laundry, if a customer's washing is finished and there is a basket available, the basket could be allocated immediately to the customer, who would then begin unloading the washing machine. Time is advanced to the next scheduled event (usually the end of an activity) and activities are examined to see whether any can now start as a consequence. Event scheduling approach has one advantage and one disadvantage as: "The advantage was that it required no specialized language or operating system support. Event-based simulations could be implemented in procedural languages of even modest capabilities." While the disadvantage "of the event-based approach was that describing a system as a collection of events obscured any sense of process flow." As such, "In complex systems, the number of events grew to a point that following the behavior of an element flowing through the system became very difficult" [35, 39, 40].

21.4.8   ACTIVITY SCANNING APPROACH

Another simulation modeling structure is activity scanning. Activity scanning is also known as the two-phase approach. Activity scanning produces a simulation program composed of independent modules waiting to be executed. In the first phase, a fixed amount of time is advanced, or scanned. In phase two, the system is updated (if an event occurs). Activity scanning is similar to rule-based programming (if the specified condition is met, then a rule is executed) [36, 40].

21.4.9 SIMULATION ASPECTS

Three aspects need to be considered when planning a computer simulation project:
1. Time-flow handling
2. Behavior of the system
3. Change handling

The flow of time in a simulation can be handled in two manners: the first is to move forward in equal time intervals. Such an approach is called time-slice. The second approach is next-event which increments time in variable amounts or moves the time from state to state. On one hand, there is less information to keep in the time-slice approach. On the other hand, the next-event approach avoids the extra checking and is more general [40, 43].

The behavior of the system can be deterministic or stochastic: deterministic system, of which its behavior would be entirely predictable, whereas, stochastic system, of which its behavior cannot be predicted but some statement can be made about how likely certain events are to occur.

The change in the system can be discrete or continuous. Variables in the model can be thought of as changing values in four ways [43]:

1. Continuously at any point of time: thus, change smoothly and values of variables at any point of time.
2. Continuously changing but only at discrete time events: values change smoothly but values accessible at predetermined time.
3. Discretely changing at any point of time: state changes are easily identified but occur at any time.
4. Discretely changing at any point of time: state changes can only occur at specified point of time.

Others define 3 and 4 as discrete event simulation as follows: "a discrete event simulation is one in which the state of a model changes at only a discrete, but possibly random, set of simulated time points." Mixed or hybrid systems with both discrete and continuous change do exist. Actually simulation packages try to include both.

In the natural sciences one makes to model the complex processes occurring in nature as accurately as possible. The first step in this direction is the description of nature. It serves to develop an appropriate system of concepts. However, in most cases, only observation is not enough to find the underlying principles. Most processes are too complex and cannot be clearly separated from other processes that interact with them. Instead, if it is possible, the scientist creates the conditions under the process is to be observed in an experiment. This method allows discovering how the observed event depends on the chosen conditions and allows inferences about the principles underlying the behavior of the observed system. The goal is the mathematical formulation of the underlying principles by using a theory of the phenomena under investigation and describes how certain variables behave independence of each other and how they change under certain conditions over time. This is mostly done by means of differential and integral equations. The resulting equations, which encode the description of the system or process, are referred to as a mathematical model.

A model that has been confirmed does not only permit the precise description of the observed processes, but also allows the prediction of the results of similar physical processes within certain bounds. Thereby, experimentation, the discovery of underlying principles from the results of measurements, and the translation of those principles into mathematical variables and equations go hand in hand. Theoretical and experimental approaches are therefore most intimately connected. The phenomena that can be investigated in this way in physics and chemistry extend over very different orders of magnitudes. They can be found from the smallest to the largest observable length scales, from the investigation of matter in quantum mechanics to the study of the shape of the universe. The occurring dimensions range from the nanometer range (10^{-9} meters) in the study of properties of matter on the molecular level to 10^{23} meters in the study of galaxy clusters. Similarly, the time scales that occur in these models (that is, the typical time intervals in which the observed phenomena take place) are vastly different. They range in the mentioned examples from 10^{-12} or even 10^{-15} seconds to 10^{17} seconds, thus from picoseconds or even femtoseconds up to time intervals of several billions of years. The masses occurring in the models are just as different, ranging between 10^{-27} kilograms for single atoms to 10^{40} kilograms for entire galaxies.

The wide range of the described phenomena shows that experiments cannot always be conducted in the desired manner. For example in astrophysics, there are only few possibilities to verify models by observations and experiments and to thereby confirm them, or in the opposite case to reject models, to falsify them. On the other hand, models that describe nature sufficiently well are often so complicated that no analytical solution can be found.

Take for example the case of the Vander Waals equation to describe dense gasses or the Boltzmann equation to describe the transport of rarefied gasses. Therefore, one usually develops a new and simplified model that is easier to solve. However, the validity of this simplified model is in general more restricted. To derive such models one often uses techniques such as averaging methods, successive approximation methods, matching methods, asymptotic analysis and homogenization. Unfortunately, many important phenomena can only be described with more complicated models. But then these theoretical models can often only be tested and verified in a few simple cases. As an example consider again planetary motion and the gravitational force acting between the planets according to Newton's law. As is known, the orbits following from Newton's law can be derived in closed form only for the two-body case. For three bodies, analytical solutions in closed form in general no longer exist. This is also true for our planetary system as well as the stars in our galaxy.

Many models, for example in materials science or in astrophysics, consist of a large number of interacting bodies (called particles), as for example atoms and molecules. In many cases the number of particles can reach several millions or more. But large numbers of particles do not only occur on a microscopic scale. These are some of the reasons why computer simulation has recently emerged as a third way in science besides the experimental and theoretical approach. Over the past years, computer simulation has become an indispensable tool for the investigation and prediction of physical and chemical processes. In this context, computer simulation means the mathematical prediction of technical or physical processes on modern computer systems [44].

The following procedure is typical in this regard: A mathematical-physical model is developed from observation. The derived equations, in most cases, valid for continuous time and space, are considered at selected discrete points in time and space. For instance, when discretizing in time, the solution of equations is no longer to be computed at all points in time, but is only considered at selected points along the time axis. Differential operators, such as derivatives with respect to time, can then be approximated by difference operators. The solution of the continuous equations is computed approximately at those selected points. The more densely those points are selected, the more accurately the solution can be approximated. Here, the rapid development of computer technology, which has led to an enormous increase in the computing speed and the memory size of computing systems, now allows simulations that are more and more realistic. The results can be interpreted with the help of appropriate visualization techniques. If corresponding results of physical experiments are available, then the results of the computer simulation can be directly compared. This leads to a verification of the results of the computer simulation or to an improvement in the applied methods or the model (for instance by appropriate changes of parameters of the model or by changing the used equations).

Altogether, for a computer experiment, one needs a mathematical model. But the solutions are now obtained approximately by computations, which are carried out by a program on a computer. This allows studying models that are significantly more complex and therefore more realistic than those accessible by analytical means. Furthermore, this allows avoiding costly experimental setups. In addition, situations can be considered that otherwise could not be realized because of technical shortcomings or because they are made impossible by their consequences. For instance, this is the case if it is hard or impossible to create the necessary conditions in the laboratory, if measurements can only be conducted under great difficulties or not at all, if experiments would take too long or would run too fast to be observable, or if the results would be difficult to interpret. In this way, computer simulation makes it possible to study phenomena not accessible before by experiment. If a reliable mathematical model is available that describes the situation at hand accurately enough, it does in general not make a difference for the computer experiment. Obviously this is different, if the experiment would actually have to be carried out in reality. Moreover, the parameters of the experiment can easily be changed. And the behavior of solutions of the mathematical model with respect to such parameters changes can be studied with relatively little effort [44] (Fig. 21.4).

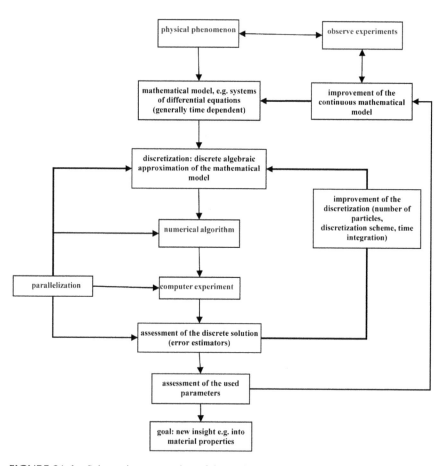

FIGURE 21.4   Schematic presentation of the typical approach for numerical simulation.

In nanotechnology numerical simulation can help to predict properties of new materials that do not yet exist in reality. And it can help to identify the most promising or suitable materials. The trend is towards virtual laboratories in which materials are designed and studied on a computer. Simulation offers the possibility of determining mean or average properties for the material macroscopic characterization. At whole it can be said, computer experiments act as a link between laboratory experiments and mathematical- physical theory.

Each of the partial steps of a computer experiment must satisfy a number of requirements. First and foremost, the mathematical model should describe reality as accurately as possible. In general, certain compromises between accuracy in the numerical solution and complexity of the mathematical model have to be accepted. In most cases, the complexity of the models leads to enormous memory and computing time requirements, especially if time-dependent phenomena are studied. Depending on the formulation of the discrete problem, several nested loops have to be executed for the time dependency, for the application of operators, or also for the treatment of nonlinearities.

Current researches therefore have focus in particular on the development of methods and algorithms that allow to compute the solutions of the discrete problem as fast as possible (multilevel and multiscale methods, multiple methods, fast Fourier transforms) and that can approximate the solution of the continuous problem with as little memory as possible. More realistic and therefore in general more complex models require faster and more powerful algorithms. Another possibility to run larger problems is the use of vector computers and parallel computers. Vector computers increase their performance by processing similar arithmetic instructions on data stored in a vector in an assembly line-like fashion. In parallel computers, several dozens in many thousands of powerful processors (the processors in use today) have mostly a RISC (reduced instruction set computer) processor architecture. They have fewer machine instructions compared to older processors, allowing a faster, assembly line like execution of the instructions which are assembled into one computing system [45]. These processors can compute concurrently and independently and can communicate with each other to improve portability of programs among parallel computers from different manufacturers and to simplify the assembly of computers of various types to a parallel computer which has uniform standards for data exchange between computers. A reduction of the required computing time for a simulation is achieved by distributing the necessary computations to several processors. Up to a certain degree, the computations can then be executed concurrently. In addition, parallel computer systems in general have a substantially larger main memory than sequential computers. Hence, larger problems can be treated [44].

Figure 21.5 shows the development of the processing speed of high performance. The performances in flop/s is plotted versus the year, for the fastest parallel computer in the world, and for the computers at position 100 and 500 in the list of the fastest parallel computers in the world. Personal computers and workstations have seen a similar development of their processing speed. Because of that, satisfactory simulations have become possible on these smaller computers [44].

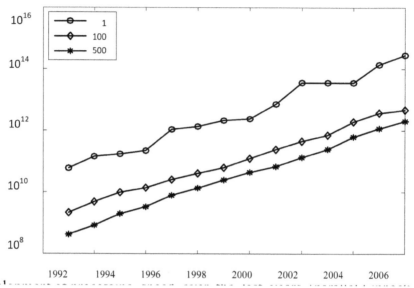

FIGURE 21.5 Development of processing speed over the last years (parallel Lin pack benchmark); fastest (1), 100th fastest (100) and 500th fastest (500) computer in the world; up to now the processing speed increases tenfold about every four years.

21.5 PARTICLE MODELS

An important area of numerical simulation deals with so-called particle models. These are simulation models in which the representation of the physical system consists of discrete particles and their interactions. For instance, systems of classical mechanics can be described by the positions, velocities, and the forces acting between the particles. In this case the particles do not have to be very small either with respect to their dimensions or with respect to their mass, as possibly suggested by their name. Rather they are regarded as fundamental building blocks of an abstract model. For this reason the particles can represent atoms or molecules. The particles carry properties of physical objects, as for example mass, position, velocity or charge. The state and the evolution of the physical system are represented by these properties of the particles and by their interactions, respectively [46]. The use of Newton's second law results in a system of ordinary differential equations of second order describing how the acceleration of any particle depends on the force acting on it. The force results from the interaction with the other particles and depends on their position. If the positions of the particles change relative to each other, then in general also the forces between the particles change. The solution of the system of ordinary differential equations for given initial values then leads to the trajectories of the particles. This is a deterministic procedure, meaning that the trajectories of the particles are in principle uniquely determined for all times by the given initial values.

But why is it reasonable at all to use the laws of classical mechanics when at least for atomic models the laws of quantum mechanics should be used? Should not Schrodinger's equation be employed as equation of motion instead of Newton's laws? And what does the expression "interaction between particles" actually mean, exactly?

If one considers a system of interacting atoms, which consists of nuclei and electrons, one can in principle determine its behavior by solving the Schrodinger equation with the appropriate Hamilton operator. However, an analytic or even numerical solution of the Schrodinger equation is only possible in a few simple special cases. Therefore, approximations have to be made. The most prominent approach is the Born-Oppenheimer approximation. It allows a separation of the equations of motions of the nuclei and of the electrons. The intuition behind this approximation is that the significantly smaller mass of the electrons permits them to adapt to the new position of the nuclei almost instantaneously. The Schrodinger equation for the nuclei is therefore replaced by Newton's law. The nuclei are then moved according to classical mechanics, but using potentials that result from the solution of the Schrodinger equation for the electrons. For the solution of this electronic Schrodinger equation approximations have to be employed. Such approximations are for instance derived with the Hartree-Fock approach or with density functional theory. This approach is known as Ab initio molecular dynamics. However, the complexity of the model and the resulting algorithms enforces a restriction of the system size to a few thousand atoms [46].

A further simplification is the use of parameterized analytical potentials that just depend on the position of the nuclei (classical molecular dynamics). The potential function itself is then determined by fitting it to the results of quantum mechanical electronic structure computations for a few representative model configurations and subsequent force-matching or by fitting to experimentally measured data. The use of these very crude approximations to the electronic potential hyper-surface allows the treatment of systems with many millions of atoms. However, in this approach quantum mechanical effects are lost to a large extent.

21.5.1 PHYSICAL SYSTEMS FOR PARTICLE MODELS

The following list gives some examples of physical systems that can be represented by particle systems in a meaningful way. They are therefore amenable to simulation by particle methods [47]:

21.5.1.1 SOLID STATE PHYSICS

The simulation of materials on an atomic scale is primarily used in the analysis of known materials and in the development of new materials. Examples for phenomena studied in solid state physics are the structure conversion in metals induced by temperature or shock, the formation of cracks initiated by pressure, shear stresses, etc. in fracture experiments, the propagation of sound waves in materials, the impact of defects in the structure of materials on their load-bearing capacity and the analysis of plastic and elastic deformations [48].

21.5.1.2   FLUID DYNAMICS

Particle simulation can serve as a new approach in the study of hydro-dynamical instabilities on the microscopic scale, as for instance, the Rayleigh-Taylor or Rayleigh-Benard instability. Furthermore, molecular dynamics simulations allow the investigation of complex fluids and fluid mixtures, as for example emulsions of oil and water, but also of crystallization and of phase transitions on the microscopic level [49].

21.5.1.3   BIOCHEMISTRY

The dynamics of macromolecules on the atomic level is one of the most prominent applications of particle methods. With such methods it is possible to simulate molecular fluids, crystals, amorphous polymers, liquid crystals, zeolites, nuclear acids, proteins, membranes and many more biochemical materials [50].

21.5.1.4   ASTROPHYSICS

In this area, simulations mostly serve to test the soundness of theoretical models. In a simulation of the formation of the large-scale structure of the universe, particles correspond to entire galaxies. In a simulation of galaxies, particles represent several hundred to thousand stars. The force acting between these particles results from the gravitational potential [46, 51].

21.5.2   COMPUTER SIMULATION OF PARTICLE MODELS

In the computer simulation of particle models, the time evolution of a system of interacting particles is determined by the integration of the equations of motion. Here, one can follow individual particles, see how they collide, repel each other, attract each other, how several particles are bound to each other, are binding to each other, or are separating from each other. Distances, angles and similar geometric quantities between several particles can also be computed and observed over time. Such measurements allow the computation of relevant macroscopic variables such as kinetic or potential energy, pressure, diffusion constants, transport coefficient, structure factors, spectral density functions, distribution functions, and many more.

In most cases, variables of interest are not computed exactly in computer simulations, but only up to certain accuracy. Because of that, it is desirable to achieve:
- an accuracy as high as possible with a given number of operations;
- a given accuracy with as few operations as possible; or
- a ratio of effort (number of operations) to achieved accuracy which is as small as possible.

Clearly the last alternative includes the first two as special cases. A good algorithm possesses a ratio of effort (costs, number of operations, necessary memory) to benefit (achieved accuracy) that is as favorable as possible. As a measure for the ranking of algorithms one can use the quotient:

$$\frac{effort}{benfit} \neq \frac{operations}{achieved\ accuracy}$$

This is a number that allows the comparison of different algorithms. If it is known how many operations are minimally needed to achieve certain accuracy, this number shows how far a given algorithm is from optimal. The minimal number of operations to achieve a given accuracy ε is called ε-complexity. The ε-complexity is thus a lower bound for the number of operations for any algorithm to achieve an accuracy of ε [52].

21.6 HIGH LEVEL ARCHITECTURE (HLA)

In 2000, HLA became an IEEE standard for distributed simulation. It consists of several federates (members of the simulation), that make up a federation (distributed simulation), work together and use a common runtime infrastructure (RTI). The RTI interface specification, together with the HLA, [53] object model template (OMT) and the HLA rules, are the key defining elements of the whole architecture [53–55].

TABLE 21.1   Continuous vs. Discrete Simulations

Continuous	Discrete
Continuously advances time and system state.	System state changes only when events occur.
Time advances in increments small enough to ensure accuracy.	Time advances from event to event.
State variables updated at each time step.	State variables updated as each event occurs.

21.7 HLA OBJECT MODELS

Whereas the interface specification is the core of the transmission system that connects different software systems, regardless of platform and language, the object model template is the language spoken over that line.

HLA has an object-oriented world-view, which is not to be confused with OOP (object-oriented programming) because it doesn't specify the methods of objects, since in the common case this is not info to be transferred between federates. This view does only define how a federate must communicate with other federates, while it doesn't consider the internal representation of each federate. So, a simulation object model (SOM) is built, which defines what kind of data federates have to exchange with each other. Furthermore, a meta-object model, the federation object model (FOM), collects all the classes defined by each participant to the federation in order to give a description of all shared information [56].

The object models, then, describe the objects chosen to represent real world, their attributes and interactions and their level of detail.

Both the SOM and the FOM are based on the OMT, that is, on a series of tables that describe every aspect of each object (Fig. 21.6). The OMT consists of the following 14 components:

1. Object model identification table
2. Object class structure table
3. Interaction class structure table
4. Attribute table
5. Parameter table
6. Dimension table
7. Time representation table
8. User-supplied tag table
9. Synchronization table
10. Transportation type table
11. Switches table
12. Data type tables
13. Notes table
14. FOM/SOM lexicon

HLA does not mandate the use of any particular FOM, however, several "reference FOMs" have been developed to promote a-priori interoperability. That is, in order to communicate, a set of federates must agree on a common FOM (among other things), and reference FOMs provide constructed FOMs that are supported by a wide variety of tools and federates [57].

Reference FOMs can be used or can be extended to add new simulation concepts that are specific to a particular federation or simulation domain. The RPR FOM (real-time platform-level reference FOM) is a reference FOM that defines HLA classes, attributes and parameters that are appropriate for real-time, platform-level simulations [58].

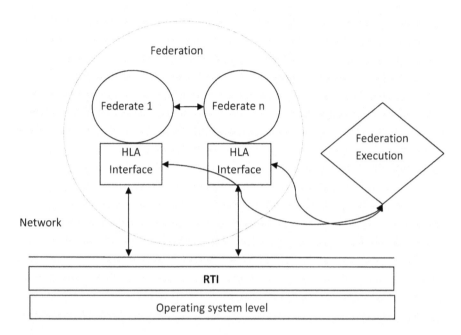

FIGURE 21.6   A distributed simulation under HLA.

21.8   OPTIMIZATION

Its objective is to select the best possible decision for a given set of circumstances without having to enumerate all of the possibilities and involves maximization or minimization as desired. In optimization decision variables are variables in the model which you have control over. Objective function is a function (mathematical model) that quantifies the quality of a solution in an optimization problem. Constraints must be considered, conditions that a solution to an optimization problem must satisfy and restrict decision variables are determined by defining relationships among them. It must be found the values of the decision variables that maximize (minimize) the objective function value, while staying within the constraints. The objective function and all constraints are linear functions (no squared terms, trigonometric functions, ratios of variables) of the decision variables [59, 60].

21.9   SIMULATION MODEL DEVELOPMENT

There are 11 steps which can expand the model simulation.
Step 1. Identify Problem
  • Enumerate problems with an existing system

- Produce requirements for a proposed system

Step 2.

a) *Formulate Problem*
  - Define overall objectives of the study and specific issues to be addressed
  - Define performance measures

b) *Quantitative criteria on the basis of which different system configurations will be evaluated and compared*
  - Develop a set of working assumptions that will form the basis for model development
  - Model boundary and scope (width of model)
  - Determines what is in the model and what is out
  - Level of detail (depth of model)
  - Specifies how in-depth one component or entity is modeled
  - Determined by the questions being asked and data availability
  - Decide the time frame of the study
  - Used for one-time or over a period of time on a regular basis

Step 3.

a) *Collect and Process Real System Data*
  - Collect data on system specifications, input variables, performance of the existing system, etc.
  - Identify sources of randomness (stochastic input variables) in the system
  - Select an appropriate input probability distribution for each stochastic input variable and estimate corresponding parameters

b) *Standard distributions*

c) *Empirical distributions*

d) *Software packages for distribution fitting*

Step 4.

a) *Formulate and Develop a Model*
  - Develop schematics and network diagrams of the system

b) *How do entities flow through the system*
  - Translate conceptual models to simulation software acceptable form
  - Verify that the simulation model executes as intended

c) *Build the model right (low-level checking)*
  - Traces
  - Vary input parameters over their acceptable ranges and check the output

Step 5.

a) *Validate Model*
  - Check whether the model satisfies or fits the intended usage of system (high-level checking)

b) *Build the right model*

- Compare the model's performance under known conditions with the performance of the real system
- Perform statistical inference tests and get the model examined by system experts
- Assess the confidence that the end user places on the model and address problems if any

Step 6. Document Model for Future Use
- Objectives, assumptions, inputs, outputs, etc.

Step 7.
a) Select Appropriate Experimental Design
- Performance measures
- Input parameters to be varied

b) Ranges and legitimate combinations
- Document experiment design

Step 8. Establish Experimental Conditions for Runs:
- Whether the system is stationary (performance measure does not change over time) or nonstationary (performance measure changes over time)
- Whether a terminating or a nonterminating simulation run is appropriate
- Starting condition
- Length of warm-up period
- Model run length
- Number of statistical replications

Step 9. Perform Simulation Runs

Step 10.
a) Analyze Data and Present Results:
- Statistics of the performance measure for each configuration of the model

b) Mean standard deviation, range, confidence intervals, etc.
- Graphical displays of output data

c) Histograms scatter plot, etc.
- Document results and conclusions

Step 11. Recommend Further Courses of Actions:
- Other performance measures
- Further experiments to increase the precision and reduce the bias of estimators
- Sensitivity analysis
- How sensitive the behavior of the model is to changes of model parameters

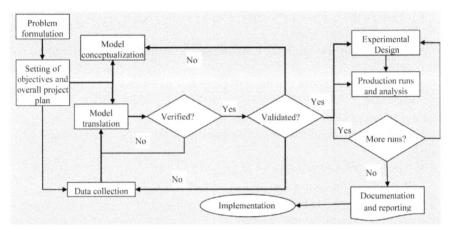

FIGURE 21.7   Steps in a simulation study.

21.10 SIMULATION LANGUAGES

SLAM introduced in 1979 by Pritsker and Pegden. SIMAN introduced in 1982 by Pegden, first language to run both a mainframe as well as a microcomputer. In primary computers, accessibility and interactions were limited. GASP IV introduced by Pritsker, Triggered a wave of diverse applications, which is significant in the evaluation of simulation.

Many programming systems have been developed, incorporating simulation languages. Some of them are general-purpose in nature, while others are designed for specific types of systems. FORTRAN, ALGOL, and PL/1 are examples of general-purpose languages, while GPSS, SIMSCRIPT, and SIMULA are examples of special simulation languages. Programming can be done in general purpose languages such as Java, simulation languages like SIMAN or use simulation packages, Arena [61, 62].

Four choices are existed: simulation language [63], general-purpose language [64], extension of general purpose [65], simulation package [66]. Simulation language is built in facilities for time steps, event scheduling, data collection, reporting. General-purpose is known to developer, available on more systems, flexible. The major difference is the cost tradeoff. Simulation language requires startup time to learn, while general purpose may require more time to add simulation flexibility. Recommendation may be for all analysts to learn one simulation language so understand those "costs" and can compare. Extension of general-purpose is collection of routines and tasks commonly used. Often, base language with extra libraries that can be called and Simulation packages allow definition of model in interactive fashion. Get results in one day. Tradeoff is in flexibility, where packages can only do what developer envisioned, but if that is what is needed then is quicker to do so.

Now some of advantages and disadvantages of common languages are listed as:

a) General Simulation Languages: Arena, Extend, GPSS, SIMSCRIPT, SIMU-LINK (Matlab), etc.

   Advantages: Standardized features in modeling, shorter development cycle for each model, Very readable code.

   Disadvantages: Higher software cost (up-front), Additional training required, limited portability.

b) Special Purpose Simulation Packages: Manufacturing (Auto Mod, FACTOR/AIM, etc.), Communications network (COMNET III, NETWORK II.5, etc.), Business (BP$IM, Process Model, etc.), Health care (Med Model).

   Advantages: Very quick development of complex models, short learning cycle, little programming.

   Disadvantages: High cost of software, limited scope of applicability, limited flexibility [67, 68].

21.11 SIMULATION CHECKLIST

You should check the simulation process in each step as:

a) Checks before developing simulation.
   - Is the goal properly specified?
   - Is detail in model appropriate for goal?
   - Does the team include the right mix (leader, modeling, programming, background)?
   - Has sufficient time been planned?

b) Checks during simulation development
   - Is random number random?
   - Is model reviewed regularly?
   - Is model documented?

c) Checks after simulation is running
   - Is simulation length appropriate?
   - Are initial transients removed?
   - Has the model been verified?
   - Has the model been validated?
   - Are there any surprising results? If yes, have they been validated?

21.12 COMMON MISTAKES IN SIMULATION

Inappropriate Level of Detail: Level of detail often potentially unlimited but more detail requires more time to develop and often to run. It can be introduced more bugs, making more inaccurate not less. Often, more detailed viewed as "better" but may not be the case. So more detail requires more knowledge of input parameters and getting input parameters wrong may lead to more inaccuracy. Therefore, start with less detail, study sensitivities and introduce detail in high impact areas.

Improper Language: Choice of language can have significant impact on time to develop, special-purpose languages can make implementation, verification and analysis easier, C++Sim, Java Sim, Sim Py (thon).

Unverified Models: Simulations generally need large computer programs unless special steps taken, bugs or errors.

Invalid Models: Unless no errors occur, the model does not represent real system, you need to validate models by analytic, measurement or intuition.

Improperly Handled Initial Conditions: Often, initial trajectory is not representative of steady state so can lead to inaccurate results. In this case, typically you want to discard, but need a method to do it so effectively.

Too Short Simulation Runs: Attempting to save time makes even more dependent upon initial conditions. Therefore, correct length depends upon the accuracy desired (confidence intervals).

Poor Random Number Generators and Seeds: "Home grown" are often not random enough to make artifacts. So the best is to use well-known one and choose seeds that are different [69, 70].

21.13   BETTER UNDERSTANDING BY A TANGIBLE EXAMPLE

21.13.1   BASIC KINETIC EQUATION

In recrystallization and transformation, a new phase forms and grows. These new phases continue to grow until they meet each other and stop growing. This situation is called hard impingement and can be expressed by using the Avrami type equation [71]:

$$X = 1 - \exp\left(-kt^n\right), \quad k = \frac{\pi \dot{N} \dot{G}^3}{3}, \quad n = 4 \tag{1}$$

Or the Johnson–Mehl equation [72]. In these equations, the concept of extended volume fraction is adopted. By using this concept, the hard impingement can be taken into consideration indirectly. The extended volume fraction is the sum of the volume fraction of all new phases without direct consideration of the hard impingement between new particles and is related to the actual volume fraction by

$$X = 1 - \exp(-X_e) \tag{2}$$

The general form of the equation was developed by Cahn [73]. A brief explanation is presented here. The nucleation sites of new phases would be grain boundaries, grain edges, and/or grain corners. In the case of grain boundary nucleation, the volume fraction of a new phase after some time can be expressed as follows. Cahn considered the situation illustrated in Fig. 21.8 and calculated the volume of the semicircle.

In his calculation, firstly, the area at the distance of y from the nucleation site B is calculated. The summation of this area for all nuclei gives the total extended area. From this value, the actual area can be calculated. The extended volume can be obtained by integrating the area for all distances. Finally, the actual volume fraction can be derived.

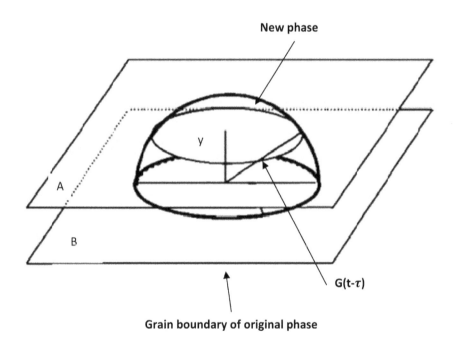

FIGURE 21.8   Schematic illustration of the situation of new phase at time τ which nucleates at time τ at grain boundary B.

The area of the section at a plane A for a semicircle nucleated at a plane B is considered. The radius r at time τ can be expressed as:

$$\text{For } y < G(t - \tau) \text{ For } y > G(t - \tau) \tag{3}$$

In this calculation, the growth rate is assumed to be constant. From this radius, the extended area fraction dY_e for the new phases nucleated at time between τ and $\tau + d\tau$ can be achieved as:

$$\text{For } y < \dot{G}(t - \tau)$$

$$\text{For } y > \dot{G}(t - \tau) \tag{4}$$

By integrating for the time τ from 0 to τ, the extended area fraction at the plane Λ at time τ can be showed as:

$$Y_e = \int_0^t dY_e = \pi I_S \int_0^{t - y/\dot{G}} \left[\dot{G}^2 (t - \tau)^2 - y^2 \right] d\tau \tag{5}$$

By exchanging $y / \dot{G}t$ with x, this equation changes to:

$$Y_e = \pi I_S \dot{G}^2 t^3 \left[\frac{1 - x^3}{3} - x^2 (1 - x) \right] \quad \text{For x<1} \tag{6}$$

$$Y_e = 0 \quad \text{For x>1}$$

The actual area fraction of new phases at plane Λ, γ can be calculated by using Y_e:

$$Y = 1 - \exp(-Y_e) \tag{7}$$

The integration of γ for y from 0 to infinity gives the volume of new phases nucleated at unit $\exp\{-\pi I_S \dot{G}^2 t^3\}$ area of plane B, V_0, as:

$$V_0 = 2\int_0^\infty Y dy = 2\dot{G}t \int_0^1 \left[1 - \exp\left\{ -\pi I_S \dot{G}^2 t^3 (\frac{1 - x^3}{3} - x^2 (1 - x)) \right\} \right] dx \tag{8}$$

Multiplying V_0 by the area of nucleation site, the extended volume fraction is obtained as:

$$X_e = SV_0 = b_s^{-\frac{1}{3}} f_s(a_s) \tag{9}$$

where

$$a_s = (I_S \dot{G}^2)^{\frac{1}{3}} t, \quad b_s = \frac{I_S}{8S^3 \dot{G}} = \frac{\dot{N}}{8S^4 \dot{G}}$$

$$f_s(a_s) = a_s \int_0^1 \left[1 - \exp\left\{ -\pi a_s^3 \left(\frac{1-x^3}{3} - x^2(1-x) \right) \right\} \right] dx \tag{10}$$

So the actual volume fraction can be expressed as:

$$X = 1 - \exp(-b_s^{-\frac{1}{3}} f_s(a_s)) \tag{11}$$

From this equation, two extreme cases can be considered. One is the case where a_s is very small and the other is extremely large. For these two cases, the equation becomes:

$$X = 1 - \exp\left(-\frac{\pi}{3\dot{N}\dot{G}^3 t^4} \right) \qquad a_s \ll 1 \tag{12}$$

$$X = 1 - \left(-2S\dot{G}t \right) \qquad a_s \gg 1 \tag{13}$$

The Eq. (12) is the same as the one obtained for the case of random nucleation sites by Johnson–Mehl. This equation implies that the increase in the volume of new phases is caused by nucleation and growth. On the other hand, Eq. (13) does not include nucleation rate and it implies that the nucleation sites are covered by new phases and the increase in the volume is dependent only on the growth of new phases. This situation is referred to as site saturation [73].

Cahn did this type of formulation for the cases of grain edge and grain corner nucleations. Table 21.2 shows all the extreme cases. For all cases, the increase of the volume of new phases for the case of small a_s conforms to the case of nucleation and growth and site saturation for the case of large a_s. The value of a_s increases when the nucleation rate is small when compared to the growth rate. The early stage of reaction corresponds to small a_s and the latter stage corresponds to large a_s. From Table 21.2, we can recognize that the exponent of time depends on the mode of reaction and the type of nucleation site for the case of site saturation. The equations in Table 21.2 can be used for calculating actual reactions such as transformation and recrystallization by introducing fitting parameters [74].

TABLE 21.2   The Kinetic Equations Depending on the Modes and the Nucleation Sites of Reaction in Accordance with Cahn's Treatment

Nucleation site	Nucleation and growth	Site saturation
Grain boundary	$$X = 1 - \exp\left(-\dfrac{\pi}{3N_s G^3 t^4}\right)$$	$X = 1 - (-2SGt)$
Grain edge		$X = 1 - \exp\left(-\pi L G^2 t^2\right)$
Grain corner		$X = 1 - \exp\left(-\left(\dfrac{4\pi}{3}\right)CG^3 t^3\right)$

21.13.2   UTILIZATION OF THERMODYNAMICS OF THE CALCULATION OF TRANSFORMATION AND RECITATION KINETICS

As transformation and precipitation kinetics are mostly related to phase equilibrium, thermodynamics can be used for their calculation. In this section, the method for using thermodynamics for the calculation will be explained.

For the consideration of kinetics, the Gibbs free-energy–composition diagram is much more useful and should be the basis. Figure 21.9 shows the Gibbs free-energy–composition diagram for austenite and ferrite in steels. Chemical composition at the phase interface between ferrite and austenite is obtained from the common tangent for free-energy curves of ferrite and austenite. The common tangent can be calculated under the condition that chemical potentials of all chemical elements in ferrite are equal to those in austenite. This condition is showed as:

$$\mu_i^\alpha = \mu_i^\gamma \tag{14}$$

In Fig. 21.9, the driving force for transformation from austenite to ferrite is indicated as well. It can be calculated by:

$$\Delta G_m = \sum x_i^\alpha (\mu_i^\gamma - \mu_i^\alpha) \tag{15}$$

These values are necessary for the calculation of moving rate of the interface during transformation and precipitation. The Zener–Hillert equation, which represents the growth rate of ferrite into austenite, is expressed as:

$$\dot{G} = \frac{1}{2r} D \frac{C_{\alpha\gamma} - C_{\gamma}}{C_{\gamma} - C_{\alpha}} \tag{16}$$

$C_{\alpha\gamma}$, Ca and C_{γ} is the carbon content in austenite, ferrite at α/γ interface and in ferrite apart from interface, respectively. The carbon content at the interface can be calculated from the common tangent between two phases as shown in Fig. 21.9. There is the other type of expression of moving rate of interface which is expressed as:

$$v = \frac{M}{V_m} \Delta G_m \tag{17}$$

The driving force in this equation can be calculated for multi component system by Eq. (15). This calculation makes it possible to consider the effect of alloying elements other than the pinning effect and the solute-drag effect [75]. Recently, some commercial software for the thermodynamic calculation have been used for this type of calculation [76].

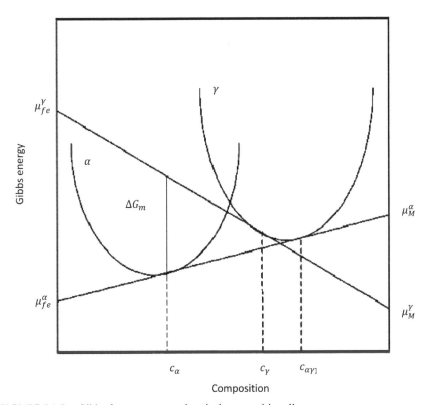

FIGURE 21.9   Gibbs free energy vs. chemical composition diagram.

21.13.3   THE CONCEPT OF THE MODEL

As mentioned in the previous section, the overall model for predicting mechanical properties of hot-rolled steels consists of several basic models: the initial state model for austenite grain size before hot-rolling, the hot-deformation model for austenitic microstructural evolution during and after hot-rolling, the transformation model for transformation during cooling subsequent to hot-rolling, and the relation between mechanical properties and microstructure of steels. In the case where steels include alloying elements, which form precipitates, the model for precipitation is essential [77].

21.13.3.1   INITIAL STATE MODEL

In this model, austenite grain sizes after slab reheating, namely before hot deformation, are calculated from the slab-reheating condition. In steels consisting of ferrite and pearlite at room temperature, austenite is formed between pearlite and ferrite and it grows into ferrite according to decomposition of pearlite. After all the microstructures become austenite, the grain growth of austenite takes place. We should formulate these metallurgical phenomena to predict austenite grain size after slab reheating. In hot-strip mill, however, the effect of initial austenite grain size on the final austenite grain size after multipass hot deformation is small. This can be due to the high total reduction in thickness by several hot-rolling steps in which the recrystallization and grain growth are repeated and the size of austenite grain becomes fine. This means that the high accuracy is not required for the prediction of the initial austenite grain size in a hot-strip mill. From this point of view, the next equation can be applied:

$$d_\gamma = exp\left\{1.61\ln\left(K + \sqrt{K^2 + 1}\right) + 5\right\} \qquad K = (T - 1413)/100 \qquad (18)$$

On the other hand, the initial austenite grain size affects the final austenite grain size in the case of plate rolling because the total thickness reduction is relatively small compared to hot-strip rolling. In this case, the high accuracy of the prediction may be needed and the model that is suitable for this case has been investigated. Three steps are considered in this model: (1) the growth of austenite between cementite and ferrite according to the dissolution of cementite, (2) the growth of austenite into ferrite at $\alpha + \gamma$ two-phase region, and (3) the growth of austenite in the γ single-phase region. The pinning effect by fine precipitates on grain growth and that of Ostwald ripening of precipitates on the grain growth of austenite are taken into consideration. This model is briefly explained in the following paragraphs.

The growth of austenite due to the dissolution of cementites can be presented as:

$$\frac{d(d_\gamma)}{dt} = \frac{D_c^\gamma}{d_\gamma} \frac{C_{\theta\gamma} - C_{\gamma\alpha}}{C_{\gamma\alpha} - C_\alpha} \tag{19}$$

$C_{\gamma\alpha}$, C_γ are the C content in austenite at γ/θ phase interface and γ/α phase interface, respectively. In the $\gamma+\theta$ two-phase region, the austenite grain size depends on the volume fraction of austenite, X_γ, which changes according to temperature. This situation is showed as:

$$d_\gamma = \left(\frac{3X_\gamma}{4\pi n_0} \right)^{1/3} \tag{20}$$

Grain growth happens in the austenite single-phase region. For grain growth, it is important to consider three cases: without precipitates, with precipitates, and with precipitates growing due to the Ostwald ripening. There are equations, which are formulated to theoretically correspond to these three cases. They are summarized by Nishizawa. The equation for the normal grain growth is expressed as:

$$d_\gamma^2 - d_{\gamma 0}^2 = k_2 t \tag{21}$$

where k_2 is the factor related to the diffusion coefficient inside the interface, the interfacial energy, and the mobility of the interface. With the pinning effect by precipitates, the growth rate becomes:

$$\frac{d(d_\gamma)}{dt} = M \left(\frac{2\sigma V}{R} - \Delta G_{pin} \right) \qquad \Delta G_{pin} = \frac{3\sigma V f}{2r} \tag{22}$$

When precipitates grow according to the Ostwald ripening, the average size of precipitates used in the Eq. (22) is obtained:

$$r^3 - r_0^3 = k_3 t \tag{23}$$

where k_3 is the factor related to temperature, interfacial energy and the diffusion coefficient of an alloying element controlling the Ostwald ripening of precipitates. By this calculation method, it is possible to predict the growth of austenite grain during heating when precipitates exist in austenite.

21.13.3.2   HOT-DEFORMATION MODEL

The hot-deformation model is required to predict the austenitic microstructure before transformation through recovery, recrystallization, and grain growth in austenitic phase region during and after multipass hot deformation.

Sellars and Whiteman [78] made the first attempt on this issue and then several researchers [79–80] developed models to calculate recovery, recrystallization, and grain growth. These models are basically similar to each other. In some models, dynamic recovery and dynamic recrystallization are taken into consideration. The dynamic recovery and recrystallization are likely to occur when the reduction is high for single-pass rolling or strain is accumulated due to multipass rolling. They should be taken into consideration in finishing rolling stands of a hot-strip mill because, the interpass time might be less than 1 sec and the accumulation of strain might take place. Here, the hot-deformation model will be explained based on the model developed by Senuma et al. [79].

In this model, dynamic recovery and recrystallization, static recovery and recrystallization, and grain growth after recrystallization are calculated as shown in Fig. 21.10. The critical strain, ϵ_c, at which dynamic recrystallization occurs is generally dependent upon strain rate, temperature, and the size of austenite grains. The effect of strain rate on ϵ_c is remarkable at low strain rate region [81].

One of the controversial issues had been whether the dynamic recrystallization took place or not when the strain rate is high such as that in a hot-strip mill. Senuma et al. [79] showed that it takes place and the effect of strain rate on ϵ_c is small at a high strain rate. The fraction dynamically recrystallized can be expressed based on the Avrami type equation as:

$$X_{dyn} = 1 - \exp\left(-0.693\left(\frac{\epsilon - \epsilon_0}{\epsilon_{0.5}}\right)^2\right) \qquad (24)$$

where $\epsilon_{0.5}$ is the strain at which the fraction dynamically recrystallized reaches 50%. On the other hand, the fraction statically recrystallized can be expressed as:

$$X_{dyn} = 1 - \exp\left(-0.693\left(\frac{t - t_0}{t_{0.5}}\right)^2\right) \qquad (25)$$

where $t_{0.5}$ is the time when the fraction statically recrystallized reaches 50% and t_0 is the starting time of static recrystallization.

The growth of grains recrystallized dynamically after hot deformation is much faster than normal grain growth in which grains grow according to square of time. This rapid growth was treated with different equations [79, 80]. The reason why this rapid growth takes place might be caused by the increase of the driving force for grain growth due to high dislocation density [79], the change in the grain boundary mobility or the annihilation of the small size grains at the initial stage. In the case of multipass deformation, the strain might not be reduced completely at the following deformation due to the insufficient time interval and the effect of accumulated strain on the recovery and recrystallization should be taken into consideration. This effect is remarkable for a hot-strip mill because of the short interpass time and for steels containing alloying elements, which retard the recovery and recrystallization. This effect can be formulated by using the change in the residual strain [82] or the dislocation density [79, 80]. In the modeling process, the accumulated strain is calculated from the average dislocation density which is obtained by calculating the changes in the dislocation density in the region dynamically recovered, ρ_n, and in the region recrystallized dynamically, ρ_s, according to time independently.

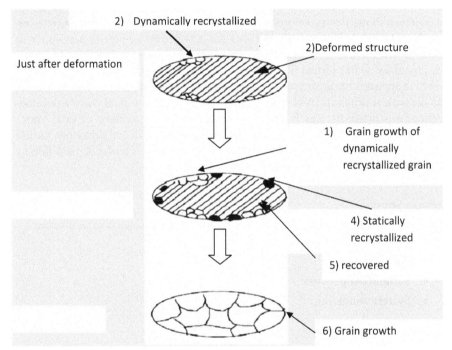

FIGURE 21.10   Schematic illustration of microstructural change due to hot deformation.

This method makes it possible to calculate the changes in grain size and dislocation density [83]. This model can be applied to the prediction of the resistance to hot deformation as well and it can contribute to the improvement of the accuracy in thickness. In this method, the average values concerning the grain size and the accumulated dislocation density are used taking the fraction recrystallized into consideration. This averaging can be applied to the hot-strip mill because the total thickness reduction is large enough to recrystallize their microstructure. In the case of plate rolling, the use of the average values is unsuitable because the reduction at each pass is small and the total thickness reduction is not enough to recrystallize the microstructure of steels. The model applicable to this case has been developed by dividing the microstructure into several groups [84, 85]

21.14 CONCLUDING REMARK

Simulation is the emulation of the operation of a real-world process or system over time. The act of simulating something first requires a model; this model represents the key characteristics or behaviors/functions of the selected physical or abstract system or process. The model represents the system itself, whereas the simulation represents the operation of the system over time. The wide range of the described phenomena shows that experiments cannot always be conducted in the desired manner. A model that has been confirmed does not only permit the precise description of the observed processes, but also allows the prediction of the results of similar physical processes within certain bounds. The simulation structure that has the greatest intuitive appeal is the process interaction method. The simulation of materials on an atomic scale is primarily used in the analysis of known materials and in the development of new materials. Key issues in simulation include acquisition of valid source information about the relevant selection of key characteristics and behaviors, the use of simplifying approximations and assumptions within the simulation, and fidelity and validity of the simulation outcomes.

KEYWORDS

- **Kinetic modeling**
- **Nanoscience**
- **Simulation process**
- **System modeling**

REFERENCES

1. Mansoori, G. A. (2005). *Principles of Nanotechnology: Molecular-based Study of Condensed Matter in Small Systems*. World Scientific. 341.
2. Wang, Z. L. (2003). New Developments in Transmission Electron Microscopy for Nanotechnology. *Advanced Materials 15(18)*, 1497–1514.
3. Schlick, T. (2010). *Molecular Modeling and Simulation: An Interdisciplinary Guide*. 2th ed. *21*, Springer. 768.
4. Turner, C. H. et al. (2008). *Simulation of Chemical Reaction Equilibria by the Reaction Ensemble* Monte Carlo Method: A Review. *Molecular Simulation, 34(2),* 119–146.
5. Starr, F. W., Schröder, T. B., & Glotzer, S. C. (2002). Molecular Dynamics Simulation of a Polymer Melt with a Nanoscopic Particle. *Macromolecules, 35(11)*: 4481–4492.
6. Pidd, M. (1998). *Computer Simulation in Management Science.*
7. Carson, J. S. (2005). *Introduction to Modeling and Simulation.* in *Simulation Conference, (2005)* Proceedings of the Winter: IEEE.
8. Joshi, U. W. et al. (1991). *Central Postage Data Communication Network*, Google Patents.

9. Paolucci, M. & Sacile, R. (2004). *Agent-Based Manufacturing and Control Systems: New Agile Manufacturing Solutions for Achieving Peak Performance.* CRC Press.

10. Holland, J. H. (1975). *Adaptation in Natural and Artificial Systems: An Introductory Analysis with Applications to Biology, Control, and Artificial Intelligence.* U Michigan Press.

11. John, H. (1992). *Holland, Adaptation in Natural and Artificial Systems.*, MIT Press, Cambridge, MA.

12. Bathe, K., Wilson, E. L. & Peterson, F. E. (1974). *SAP IV: A Structural Analysis Program for Static and Dynamic Response of Linear Systems. 73*: College of Engineering, University of California Berkeley.

13. Kostami, V., & Ward, A. R. (2010). *Analysis and Comparison of Inventory Systems: Dynamic vs Static Policies,* 1–40.

14. Luyben, W. L. (1993). *Dynamics and Control of Recycle Systems. 1. Simple Open-Loop and Closed-Loop Systems.* Industrial & engineering chemistry research, *32(3),* 466–475.

15. Ghosh, (2004). *Control Systems: Theory And Applications.*: Pearson Education. 628.

16. Huang, B. K. (1994). *Computer Simulation Analysis of Biological and Agricultural Systems*: Taylor & Francis. 880.

17. Karnopp, D. C. Margolis, D. L. & Rosenberg, R. C. (2012). *System Dynamics: Modeling, Simulation and Control of Mechatronic Systems.* Wiley.

18. Wellstead, P. E. (1979). *Introduction to Physical System Modelling.* Academic Press London.

19. Polderman, J. W., & Willems, J. C. (1998). *Introduction to Mathematical Systems Theory: A Behavioral Approach.* Springer.

20. Flash, T. & Hogan, N. (1985). The Coordination of Arm Movements: An Experimentally Confirmed Mathematical Model. *The journal of Neuroscience, 5(7),* 1688–1703.

21. Érdi, P. (1989). *Mathematical Models of Chemical Reactions: Theory and Applications of Deterministic and Stochastic Models.* Manchester University Press.

22. Gaspard, J. (1991). Deterministic and Stochastic Models. *Solid State Phenomena, 3,* 97–107.

23. Davila, J. C. & Sanmargo, M. E. (1966). An Analysis of the Fit of Mathematical Models Applicable to the Measurement of Left Ventricular Volume. *The American journal of cardiology, 18(1),* 31–42.

24. Bettonvil, B., & Kleijnen, J. P. (1997). Searching for Important Factors in Simulation Models with many Factors: Sequential Bifurcation. *European Journal of Operational Research, 96(1),* 180–194.

25. Kroese, D. P., Taimre, T., & Botev, Z. I. (2013). *Handbook of Monte Carlo Methods.* Wiley. 768.

26. Rubinstein, R. Y., & Kroese, D. P. (2011). *Simulation and the Monte Carlo Method.*: Wiley. 372.

27. Paul, R.J. & Balmer, D. W. (1993). *Simulation Modeling.* Chartwell-Bratt.

28. Banks, J. (1999). Introduction to Simulation. in *Simulation Conference Proceedings,* Winter. (1999): IEEE.

29. Banks, J. (2000*). Introduction to Simulation.* in *Simulation Conference, 2000. Proceedings.* Winter: IEEE.

30. Maria, A. (1997). Introduction to Modeling and Simulation. in *Proceedings of the 29th conference on Winter simulation.*: IEEE Computer Society.

31. Drappa, A., & Ludewig, J. (2000). Simulation in Software Engineering Training. in *Proceedings of the 22nd International Conference on Software Engineering*. ACM.

32. Pritsker, M. (2006). The Hidden Dangers of Historical Simulation. *Journal of Banking & Finance, 30(2)*: 561–582.

33. Kayes, A. B., Kayes, D. C., & Kolb, D. A. (2005). *Experiential Learning in Teams.* Simulation & Gaming, *36(3)*: 330–354.

34. Shachter, R. D. & Peot, M. A. (2013). *Simulation Approaches to General Probabilistic Inference on Belief Networks.* arXiv preprint arXiv:1304.1526.

35. Zhou, M. C. (1998). *Modeling, Analysis, Simulation, Scheduling, and Control of Semiconductor Manufacturing Systems: A Petri Net Approach.* Semiconductor Manufacturing, IEEE Transactions. *11(3)*, 333–357.

36. Martinez, J. C. & Ioannou, P. G. *Advantages of the Activity Scanning Approach in the Modeling of Complex Construction Processes.* in *Simulation Conference Proceedings, (1995). Winter.* (1995): IEEE.

37. Martinez, J. C. & Ioannou, P. G. (1999). *General-Purpose Systems for Effective Construction Simulation.* Journal of construction engineering and management, **125**(4), p. 265–276.

38. Pidd, M. (1995). *Object-Orientation, Discrete Simulation and the Three-Phase Approach.* Journal of the Operational Research Society, 362–374.

39. Testi, A., Tanfani, E., & Torre, G. (2007). *A Three-Phase Approach for Operating Theatre Schedules.* Health Care Management Science, *10(2)*, 163–172.

40. Abu-Taieh, E. M. O. (2008). *Computer Simulation Using Excel Without Programming.*: Universal-Publishers.

41. Henriksen, J. O. (1981). *GPSS-Finding the Appropriate World-View.* in *Proceedings of the 13th Conference on Winter Simulation 2.* IEEE Press.

42. Schriber, T. J. et al. (2001). *Panel: GPSS 40th Anniversary: GPSS turns 40: Selected Perspectives.* in *Proceedings of the 33nd Conference on Winter Simulation*: IEEE Computer Society.

43. Pidd, M., & Cassel, R. A. (1998). *Three Phase Simulation in Java.* in *Proceedings of the 30th Conference on Winter Simulation*: IEEE Computer Society Press.

44. Griebel, M., Knapek, S., & Zumbusch, G. (2010). *Numerical Simulation in Molecular Dynamics: Numerics, Algorithms, Parallelization, Applications*: Springer.

45. Peter, C., Daura, X., & van Gunsteren, W. F. (2000). *Peptides of Aminoxy Acids: A Molecular Dynamics Simulation Study of Conformational Equilibria under Various Conditions.* Journal of the American Chemical Society, *122(31)*, 7461–7466.

46. Sheikh, A. A. R. E., Ajeeli, A. A., & Abu-Taieh, E. M. O. (2008). *Simulation and Modeling: Current Technologies and Applications*: IGI Global Snippet.

47. Gould, H. et al. (1988). *An Introduction to Computer Simulation Methods: Applications to Physical Systems. 1*: Addison Wesley Reading.

48. Kittel, C., & McEuen, P. (1986). *Introduction to Solid State Physics. 8*, Wiley: New York.

49. Cannon, R. H. & Truxal, J. G. (1967). *Dynamics of Physical Systems.* McGraw-Hill: New York.

50. Bull, H. B. (1943). *Physical Biochemistry.*

51. Van Holde, K. E., Johnson, W., & Ho, P. S. (2006). *Principles of Physical Biochemistry.*

52. Traub, J., Wasilkowski, G., & Wozniakowski, H. (1988). *Information-Based Complexity,* Academic Press, New York.

53. Dahmann, J. S. (1997). *High Lvel Achitecture for Simulation.* in *Distributed Interactive Simulation and Real Time Applications, (1997)., First International Workshop on IEEE.*

54. Kuhl, F., Weatherly, R., & Dahmann, J. (1999). *Creating Computer Simulation Systems: An Introduction to the High Level Architecture*: Prentice Hall PTR.

55. Dahmann, J. S., & Morse, K. L. (1998). *High Level Architecture for Simulation: An Update.* in *Distributed Interactive Simulation and Real-Time Applications. Proceedings. 2nd International Workshop on. IEEE.*

56. Nance, R. E., & Sargent, R. G. (2002). *Perspectives on the Evolution of Simulation.* Operations Research, *50(1):* 161–172.

57. Zalcman, L., & Blacklock, J. (2010). *USAF Distributed Mission Operations, an ADF Synthetic Range Interoperability Model and an AOD Mission Training Centre Capability Concept Demonstrator-What Are They and Why Does RAAF Need Them?*

58. Zeigler, B. P. & Lee, J. S. (1998). *Theory of Quantized Systems: Formal Basis for DEVS/ HLA Distributed Simulation Environment.* in *Aerospace/Defense Sensing and Controls.*: International Society for Optics and Photonics.

59. Law, A. M., & McComas, M. G. (2002). *Simulation Optimization: Simulation-Based Optimization.* in *Proceedings of the 34th Conference on Winter Simulation: Exploring New Frontiers.*: Winter Simulation Conference.

60. Banks, J. (1998). *Handbook of Simulation: Principles, Methodology, Advances, Applications, and Practice.* John Wiley & Sons. 849.

61. Eldredge, D. L., McGregor, J. D., & Summers, M. K. (1990). Applying the Object-Oriented Paradigm to Discrete Event Simulations using the C++ Language. *Simulation. 54(2),* 83–91.

62. Tocher, K. D. (1965). Review of Simulation Languages. Journal of the Operational Research Society. *16(2),* 189–217.

63. Healy, K. J. & Kilgore, R. A. (1997). *Silk: A Java-Based Process Simulation Language.* in *Proceedings of the 29th conference on Winter simulation.* IEEE Computer Society.

64. Sauro, H. M. (1993). *SCAMP: A General-Purpose Simulator and Metabolic Control Analysis Program.* Computer applications in the biosciences: CABIOS, *9(4):* 441–450.

65. Sherwood, P. et al. (2003). QUASI: A General Purpose Implementation of the QM/MM Approach and its Application to Problems in Catalysis. *Journal of Molecular Structure: THEOCHEM, 632(1),* 1–28.

66. Smith, W., & Forester, T. R. (1996). DL_POLY_2. 0: A General-Purpose Parallel Molecular Dynamics Simulation Package. *Journal of Molecular Graphics, 14(3),* 136–141.

67. Gray, M. A. (2007). Discrete Event Simulation: A Review of SimEvents. Computing in Science & Engineering, *9(6),* 62–66.

68. Huntsinger, R. C. (2003). *Simulation Languages and Applications,* in *Modeling and Simulation: Theory and Practice,* Springer. 145–154.

69. Jain, R. (1991). *The Art of Computer Systems Performance Analysis: Techniques for Experimental Design, Measurement, Simulation, and Modeling* Wiley. 685.

70. Yilmaz, L. & Ören, T. (2009). *Agent Directed Simulation and Systems Engineering.* John Wiley & Sons. 600.

71. Yang, J., McCoy, B. J. & Madras, G. (2005). Distribution Kinetics of Polymer Crystallization and the Avrami Equation. *Journal of Chemical Physics, 122(6):* 064901.

72. Johnson, W. A., & Mehl, R. F. (1939). *Reaction Kinetics in Processes of Nucleation and Growth.* Trans. Aime, *135(8),* 396–415.

73. Cahn, J. W. (1956). *Transformation Kinetics During Continuous Cooling.* Acta Metallurgica, *4(6),* 572–575.

74. Umemoto, M. (1990). *Proceedings of International Symposium on Mathematical Modelling of Hot Rolling of Steel.* Yue, S., Ed.; CIM, Quebec, 404.

75. Kaufman, L. & Vernstein, H. (1970). *Computer Calculation of Phase Diagrams.* Academic Press, New York.

76. Sundman, B.o., Jansson, B.o. & Andersson, J. O. (1985). *The Thermo-Calc Databank System.* Calphad, *9(2),* 153–190.

77. Suehiro, M., et al. (1988). *Proceedings of International Conference on Physical Metallurgy of Thermomechanical Processing of Steelsand Other Metals* ISIJ, 791.

78. Sellars, C. M. & Whiteman, J. A. (1979). Recrystallization and Grain Growth in Hot Rolling. *Metal Science, 13(3–4),* 187–194.

79. Senuma, T., et al. (1984). Structure of Austenite of Carbon-Steels in High-Speed Hot-Working Processes. Tetsu to Hagane. *Journal of the Iron and Steel Institute of Japan, 70(15),* 2112–2119.

80. Yoshie, A., et al. (1987). *Formulation of Static Recrystallization of Austenite in Hot Rolling Process of Steel Plate.* Transactions of the Iron and Steel Institute of Japan, *27(6),* 425–431.

81. Sakai, T. & Jonas, J. J. (1984). *Overview no. 35 Dynamic Recrystallization, Mechanical and Microstructural Considerations.* Acta Metallurgica, *32(2),* 189–209.

82. Saito, Y., Enami, T., & Tanaka, T. (1985). *The Mathematical Model of hot Deformation Resistance with Reference to Microstructural Changes During Rolling in Plate Mill.* Transactions of the Iron and Steel Institute of Japan, *25(11),* 1146–1155.

83. Suehiro, M., et al. (1987). *Computer Modeling of Microstructural Change and Strength of Low Carbon Steel in Hot Strip Rolling.* Transactions of the Iron and Steel Institute of Japan, *27(6),* 439–445.

84. Tomota, Y. et al. (1992). *Prediction of Mechanical Properties of Multi-phase Steels Based on Stress-Strain Curves.* ISIJ international, *32(3),* 343–349.

85. Raabe, D. et al. (2006). *Continuum Scale Simulation of Engineering Materials, Fundamentals-Microstructures Process Applications,* John Wiley & Sons.

INDEX

Milton Keynes UK
Ingram Content Group UK Ltd.
UKHW031144141024
449569UK00024B/1082